Threads of Rebellion

Book One Of The Threadbreaker Series

Brian J. Bork

For permission requests, write to the author, addressed to:

Brian J. Bork

info@bjb-author.com

Dedication

To my son, Justin, whose laughter once filled rooms like sunlight through stained glass, who was, and forever will be, Theryn to my Edrin. In the quiet moments between chapters, the memory of his wit taught mine how to build worlds neither of us could have imagined alone. He taught me more than I taught him—about courage in the face of Morthos' touch, and about the weight of doing what is needed when it is needed.

TABLE OF CONTENTS

Preface

The Loom does not speak, but it can sing, the click and clack of the treadles and the hiss of the shuttle. It was birthed, silently a rupture at consciousness's edge, a void that devoured existence before the first syllable gasped into being. Unwitnessed and primal, the Loom tore through nothingness long before concepts like 'thread' or 'destiny' could be conceived, before reality itself had bones to break. It learns to sing as a discordant symphony. Its frequencies slice through oblivion's membrane. Its notes hemorrhage from collective terror, bestial hunger, and the ash of civilizations that never were. Before gods crawled from the cosmic womb, before hope or despair had flesh to inhabit, the Loom simply shaped. Neither creator nor destroyer, but the compulsion to penetrate, to shape, to mold creation into order. It writhed with blind purpose, its tendrils coiling through the primordial void like silver-black serpents, glistening with the sweat of potential. Where these tendrils intersected, reality crystallized first as motes of quantum dust, then as churning nebulae bleeding color into the abyss. The Loom's pulsations sent ripples through this nascent matter, compressing chaos into the first trembling laws of physics, squeezing raw existence until it wept the elements that would become stars, planets, and eventually, flesh.

The Loom began to weave the world of Aurelia, its silver-black shuttles crashing through the cosmic void with terrible precision. Every pass of the world around its molten sun became another threadturn, each filament glistening with the raw potential of days yet unlived. Each turn began with Firstweave—

threads of pale dawn-gold trembling with frost then darkened and thickened into seasons: fertile spirals of spring growth, the heavy crimson weft of summer heat, autumn's brittle copper strands. The pattern grew more intricate with each pass, spinning weather, migration, and harvest into existence, until the weave somehow resolved to Newthread those gossamer filaments of rebirth glimmering like ice crystals in starlight. And so it went, the Loom's massive needles puncturing reality itself, bleeding time into being, stitching the brutal cycle of seasons across the writhing flesh of the world.

From the Loom's woven tapestry emerged life in all its forms beasts of field and forest, creatures of sea and sky, and finally mortals with their fragile, yearning hearts. As these humans gazed into the abyss of their own insignificance, their collective terror and wonder crystallized into divine form. The gods manifested not as creators, but as living echoes of mortal longing—each deity reflecting humanity's deepest hungers in them. When the first tribe huddled against winter's teeth, their terror of freezing darkness conjured Pyrrhos from their desperate flames. They fed him their nightmares of forest fires that devoured villages whole, their awe at lightning strikes that cleaved ancient trees. His laughter became the crackle of their hearths, his rage the unpredictable hunger consuming both enemies and friends. Thaliora emerged from fishermen's prayers and drowning sailors' last gasps; every tide that returned beloved boats or swallowed them became her silver hair, her mercy, and her cruelty inseparable. Umbryos coalesced from whispers exchanged in shadows, from the fear of what lurks beyond firelight's edge—assassins' blades, lovers' secrets, children's monsters given substance through belief. Each deity stumbled into being through mortal observation, naked until clothed in ritual and story. Auron's voice rose from thunder that shook valley-dwellers to their knees; Liora from the wonder of rainbows after storms; Gaelora from mountain-folk who felt the earth's grief in landslides; Morthos from every deathbed vigil where the living witnessed that final, mysterious pause.

But gods never stay the same. They writhe within their newborn forms, ravenous for definition. Their hungers carve new valleys across continents; their jealousies birth oceans. Their first childish tantrums shatter mountain ranges. Their laughter plants forests. Each divine breath rewrites them, each mortal prayer reshapes them, until the gods who end epochs bear only the faintest shadow-memory of those who began them. And the gods need one thing above all else: belief—that sacred nectar that flows from mortal minds, sweeter than ambrosia and more potent than any elixir. Without it, their magnificence withers like flowers in frost, their voices fade to whispers, and their very substance dissolves into the cosmic void.

For a time, the Loom pulsed quietly beneath their hands, each god tracing its grooves and channels, learning to coax shape from the murk. They cut their first glyphs into the fabric of Aurelia—sacred geometries capturing raw concepts like 'growth' and 'decay' in their angles, symbols that bound chaos into order through their very shape. These marks, neither language nor art but something more primal, resonated with the Loom's frequencies, translating divine will into physical law. Once carved, each glyph became a conduit through which reality's skeleton could form and harden. They quarried belief from the void, hammered it into mountain spines, river roars, and the fragile rhythm of mortal blood. Humanity, born of the Loom's overflow, delighted them with endless contradiction: gentle and cruel, wise and oblivious, so easily broken yet swift to heal. With each rhythmic thrust and pull, the Loom's mechanical cadence became the pulse beneath all living things—the silent conductor of existence itself.

But even the Loom's children could not predict the first broken weave.

It began as a shiver in the filaments, an itch in the song. The gods, drunk on power and experimentation with the mortals, noticed only too late: a whisper where there should have been silence, a quiver in the pattern that was neither divine nor mortal but something else entirely. A mortal—her name lost and forbidden—reached up and plucked her own thread from the weave. She did not beg or bargain; she simply acted, and the Loom's tapestry convulsed in shock. The gods themselves had never contemplated trying what this mortal had. For a moment, every god felt their own shape flicker, their names threatened with erasure.

In that singular act of defiance, she ripped reality's fabric—a gash in the cosmic weave raw and gaping to this day. Through this wound, choice itself hemorrhaged into Aurelia's veins. Such a simple power—to believe or disbelieve—became her gift and her sentence. Gods snuffed her rebellion quickly, as they always had. Its aftershock pulses, though, still in the bloodline of every human since born. The Loom kept weaving, for it has no other purpose—silver-black threads snapping like brittle bones, destinies tangling into grotesque knots pulsing with sickly light. Its massive shuttles tore through the warp and weft of reality, each pass leaving frayed edges where possibility bled like ichor. And when threads flayed freely, their raw ends whipping through the cosmic void, chaos gained a little more ground on order—a creeping rot of formlessness devouring pattern wherever the weave grew thin.

Startled by this sudden vulnerability, the gods turned on one another. Some whispered encouragement at the edge of mortal dreams, fanning sparks of doubt and hope. Others hissed warnings that scorched the sky. When words failed, they waged a war of will, setting the Loom's substance afire. Belief became a currency more precious than gold or blood. Each mortal's faith glimmered like dewdrops in the cosmic web, tiny luminous pearls that gods jealously collected, each droplet containing the power to reshape mountains or birth stars when properly channeled through divine will.

Thus erupted the Mantle Wars—not mere conflict, but reality's violent seizure. The cosmos itself convulsed as gods, mad with betrayal, tore at creation's fabric, tearing through the woven tapestry of Aurelia. Pyrrhos' rage ignited entire continents; Umbryos' spite drowned cities in shadow-tide. Across Aurelia, mountains liquefied and resolidified in heartbeats, oceans boiled away to reveal writhing abyssal horrors, and the night sky hemorrhaged stars that screamed as they fell. Mortals woke vomiting prophecies not their own, children were born with eyes that reflected battles yet to come, and the dead clawed up through soil to deliver warnings in voices of ash and stone. Every breath drawn became allegiance declared. Every silent prayer triggered divine retribution. Every moment of doubt opened flesh to possession.

By the time that the Weave Concord formed, ending the Mantle Wars, the Loom's suffering birthed fragmentation where once wholeness reigned. Reality shattered like struck crystal—the divine realm shimmering above, where gods nursed their wounds and plotted. The mortal realm below, where humans rebuilt from cosmic wreckage. And between them, the hidden twilight territories where half-gods, fey creatures, and heroes wandered through mist and memory. Yet the Loom itself remained apart from all its creations, neither above nor below—invisible except to those whose perception had been broken and remade by knowledge. In its fractured state, the Loom continued its ancient vigil, not merely weaving or unraveling, but standing sentinel at reality's edge, a wounded guardian whose protection Aurelia would one day recognize when all other defenses failed.

When the last embers of the wars cooled in the ash of what was left of the world, the Loom still throbbed in darkness, but its song changed forever. Seven gods circled each other warily, mantles scorched and patched, uncertain if they were jailers or prisoners of the fate they had sewn. Ready at a moment's notice, to resume their hostilities.

Pyrrhos, pride undimmed, stalked the horizon hunting rebellion's spark.

Thaliora, haunted by what might have been, drifted through mortal dreams, her tears falling as salt rain.

Umbryos, born in schism's chaos, slipped between shadows cataloguing secrets for vengeance as yet unnamed.

Gaelora seeded silent worlds, hoping something might grow beyond the Loom's pain.

Liora hid from the others and looked for the sins and confessions of the world, even as her own hands trembled with the urge to cut threads.

Auron, battered but unbroken, inhaled the new world's chaos and exhaled it as storms, his laughter edged with promise of freedom.

Morthos, last and least, watched the patterns ravel and unravel, loyal only to the quiet that must eventually claim all things.

The Loom's new rhythm beat beneath their discord—a wild, unstoppable pulse no god could silence. Gathered at sense's edge, every heartbeat a challenge, every pause a chance for treachery. They conspired and slashed, each desperate to define the next pattern. Yet none could forget that forbidden act—a single mortal, daring to will her own fate, forced the Loom to remember her forever in its weave.

Now the Loom gazes upon what remains: the fractured kingdoms spread across the continents—from Emberfield's ash-dusted plains to Veilwood's shimmering hollows, from the towering spires of Mantle Crown to the whisper-filled markets of Threadmere. It watches humans rebuild their broken cities, stone by stone, while scaled Lizardkin bask, studying scrolls on sun-warmed ruins. The Fairykin retreat to twilight territories as the catfolk hunters prowl the borderlands, their tufted ears alert for danger. Half-gods walk among mortals disguised, and high above them all, Aerithari messengers ride thermal currents, their feathered wings catching the last light of day. The Loom sees them all—each carrying the ember of that first defiance.

Now, in this fractured present, the Loom pulses again—curious, anticipating either a new defiance or a return to order. The gods sense a disturbance rising, raw and wild, in mortal blood and breath. Each prepares for the coming storm, unsure if it will destroy them or set them free at last from the Loom.

The Loom pulses once more—its song louder, more erratic.

CHAPTER ONE

HALORETH'S ARCHIVE

Chapter One

Haloreth's Archive – Twilight
(*Commonsday, 1 Dawnskein, TT 1215*)

The sealed stacks of Haloreth's Archive slumbered beneath a blanket of eternal dusk, immune to the passage of centuries or the shifting of seasons. The uninitiated imagined the place as an ossuary for dead knowledge—a maze of brittle scrolls and crumbling treatises. But those few with the will and wits to breach its defenses knew better. Here, in the limestone marrow of one of the oldest temples, the records of creation were chained with wards more insidious than any iron. Each shelf a sentry, every volume a landmine of memory waiting for the fool who dared trespass.

The mirror mage moved through this labyrinth like a shadow from a half-remembered nightmare. Her silent footfalls were buffered by years of practiced anticipation. It was not silence born of magic—such things left echoes—but of supreme discipline, a kind of negative presence that starved the world of her passage. Her figure was mutable as smoke depending on how one squinted, but in the Archive's unmoving gloom, she became elemental: all hunger, all intent.

Dust motes danced in the high shafts of light slicing down from barred windows, catching on the edges of the mirror witch's silver-blue hair. Here, the air vibrated with the static charge of unshed revelation; even the mildew held its breath, unwilling to infect anything so carefully preserved. She counted sections as she moved, lips pursed in silent arithmetic, her right hand hovering just above the row of catalog markers embedded at shoulder height. There was no need to double-check—she could have navigated the Archive blindfolded. But caution was her second nature, grafted on by years of surviving the Mirrathar family's tutelage and the more lethal politics of the Mantle Council.

Eighteen, nineteen, twenty. She turned left at the final junction, into a forgotten alcove shrouded in the sickly green glow of half-powered glyph lanterns. This was where the forbidden began—the place where even the Archive Keepers dared not linger, save on orders from the gods themselves. The illusionist paused at the alcove's mouth, head tilted just so, parsing the air for the faintest ripple of movement. Her ears caught only the distant scritching of pens from the main scriptorium; no breathing, no shuffling of robed feet, not even the scurrying of Archive mice. Perfect.

Now for the wards.

Most would have assumed the entrance was guarded by brute force, locked and braced with celestial iron, or etched with runes that howled at the touch. But the Archive favored subtler poisons—three wards in sequence, each keyed to a different spectrum of intention. The first, an aversion field, coaxed the will to turn back. The second, a truth-binder, laid bare any false purpose in the trespasser's heart. The last—newly added, and only recently discovered by the mage—was a mnemonic snare designed to wipe the memory of anyone who succeeded at the first two.

She closed her eyes, breathed once through her mouth, and began.

With her left hand, she traced a sigil in the air, index and middle fingers forming a triangle, then a spiral, then a sharp flick downward, shattering the invisible field in front of her. The aversion broke like a soap bubble, unnoticed by all except those with the means to see.

For the truth-binder, the mirror witch slowed her pulse, recalling the Mirrathar family precept: *If you must lie, first convince yourself.* She conjured a memory of her mother's disappointment, the look that had always followed any hint of imperfection. *I seek only to study, to serve the greater good, to preserve the Archive's legacy for all who come after.* The words rang true in her chest, resonating for just long enough to slip past the ward's scrutiny.

The mnemonic snare was more delicate. She felt its tendrils licking at her consciousness, hunting for a loose end to unravel. She bared her teeth in a grin, savoring the moment's challenge. She plucked a thin obsidian charm from her sleeve, rolled it between thumb and forefinger, and muttered a phrase so soft it barely disturbed the air: "Let none who see remember." The charm crumbled into dust, and the mnemonic snare recoiled, its appetite blunted.

She was through.

The alcove's heart was a pedestal of charred wood, holding a single scroll case, wrapped in bronze wire and sealed with a glyph that shimmered with active malice. The mirror witch eyed the lock, fingers twitching with anticipation. No ordinary archivist would risk it; no lesser mage would survive. Yet here she stood, alive and alone, a heartbeat away from history's most forbidden secret.

"Now, now, little ward," she whispered, her voice a velvet thread. "Show me your worst."

She touched the wire. It bit her thumb, drawing a bead of blood that hissed as it touched the glyph. For a second, nothing happened. Then the glyph blinked twice, unspooled like a dying spider, and the preservation spell hissed its last, surrendering the scroll case to mortal hands for the first time in decades.

The sorceress lifted the case, surprised by its weight: heavier than it should have been. The burden of what it contained was already tugging at her mind. She uncapped it and let the scroll slide onto the pedestal, unrolling with ponderous reluctance.

The pattern inked onto the vellum was as beautiful as it was obscene. An illusion-glyph, yes, but of a sophistication that defied everything taught in the Mirrathar halls. It did not merely hide its bearer. It rewrote the rules by which the world perceived them, layering deception atop reality in a recursive spiral, threatening to collapse under its own audacity. The edges shimmered in the low light, shifting from silver to deep violet and back again.

The ritualist's hands trembled, just barely, as she hovered her fingers above the diagram. Her mind raced to absorb every nuance: the intersection of curves, the places where the ink doubled back on itself, the tiny deviations that would have ruined the effect if traced by anyone less obsessive.

A laugh—high, involuntary, almost childish—escaped her lips. She clapped her hand over her mouth, then scanned the alcove for any sign she'd been overheard. Nothing. The air remained dead, the silence absolute.

She dared another look at the glyph. This time, she traced its form with her eyes alone, mapping it onto the endless memory palace she'd built in childhood to survive her family's lessons. She could feel the pattern wanting to sink into her skin, to nest in the ridges of her brain and reside there forever.

As she memorized the last of the interlocking sigils, the watcher allowed herself the smallest smile—a curve of the lips of a revolutionary, or a god. She replaced the scroll, wiped the pedestal clean of blood and dust, and re-wrapped the case in its wire, hands steady now with the knowledge of what she'd done.

She turned, readying herself for the return journey through the Archive, mind already whirring with the next steps. The world outside would remain unchanged, for now, but beneath its surface, reality's seams had begun to loosen.

#

There was a space in the Archive reserved for work that could not be permitted, even in theory. The shelves here hunched closer together, forming a narrow aisle choked by centuries of ignored revision and suppressed history. At the aisle's end, someone—likely an ancestor or equally desperate predecessor—had built a makeshift workbench from the bones of discarded shelving and the lid of a defunct catalog drawer. It was here, in the Archive's ossuary of failed secrets, that the mirror witch established her operation.

She set her implements on the battered plank: etching stylus, sanded cloth, crucible and tongs, whetstone, and finally the object itself—a pendant no wider than a thumbnail, mirror-surfaced and framed in a filigree of old silver. The mirrored disk looked unimpressive, its brilliance dulled by tarnish and the thumb-smears of previous owners. To the casual observer, it was a bauble; to the mirror-worker, it was an uncut diamond, the heart of a forbidden star.

Her hands worked without hesitation. She'd drilled herself on the Mirrathar tradition of preparation. Every step rehearsed in darkness, eyes closed, senses starved until the only thing that existed was the pulse of intent in her fingertips. But it was something different to perform the act under the weight of the Archive's surveillance, even if every spell and counterspell had been deployed. Shadows shifted on the workbench as the mirror witch laid out her instruments, and the lamp-glow, filtered through years of grease and dust, cast everything in a colorless pallor.

She began by polishing the mirror's reverse side. The cloth rasped softly, then fell silent as the layer of oxidation vanished, revealing a plane of unblemished, virginal silver. In that empty surface, the mirror witch glimpsed her own face: not as others saw her, but as a fragmented mask, the eyes replaced by twin points of blue-violet focus. She made a mental note to eat something—her cheeks looked sallow, and her mouth was drawn with the tension she'd sworn to avoid.

The stylus came next. She rolled it between her thumb and index finger, warming it with the friction of her anticipation, then dipped the tip into a minute vial of acid solution. This was the Mirrathar way: no ink, no paint—only the corrosive bite of etching, so that each line was a scar, a permanent trespass against the material. The practitioner touched the stylus to the mirror and pulled, slow and even, letting the tool score the first curve of the glyph.

The forbidden pattern was a spiral, but not coiling inward toward comfort or understanding. It doubled back, stuttered, and then jutted into a vector, threatening to split the pendant in half. The act of drawing it was a physical revulsion, a wrongness that buzzed up through her wrist and elbow into the base of her skull. The mirror mage gritted her teeth and continued. Each segment had to be perfect; the glyph's power depended on the absolute precision of the line, no gaps or hesitations permitted. A single slip and the illusion would shatter. Worse, it might rupture the boundary between the pendant and the attention of the gods themselves.

She completed the first spiral, then set the stylus aside to breathe. Her pulse hammered in her neck, and the world seemed to tilt slightly, as if the Archive's sub-basement had rotated by a degree or two while she worked. The illusionist pressed her palm to the cool surface of the table, grounding herself, and then, with a grimace, resumed the next layer.

The second line was thinner, so fine that she had to brace her pinkie against the mirror's edge and flex the stylus with only the weight of her breath. This line wove through the first, sometimes kissing its edge, sometimes running alongside it at a distance calculated in fractions of a hair. Her father's voice drifted through her memory, reciting the ancient instruction: *A glyph is not a word, not a command, but a knife to cut the pattern of the world. Hold steady, and let the world bleed for you.*

The smell of acid and hot metal rose from the pendant as she worked, mingling with the staleness of paper rot and the distant tang of mildew. By the time she finished the final turn, the mage's knuckles had gone white, and her back ached from the tension of holding so perfectly still.

She laid down the stylus and, for a moment, simply watched as the glyph began to manifest. At first, it looked like nothing—an arcane doodle, a student's attempt at forbidden graffiti. Then the lines shimmered, a slow iridescence blooming along the etch. The mirror's surface warped, refusing to settle on a single reflection. As she watched, her face fractured and multiplied, resolving into an unfamiliar, featureless mask. She reached out and touched the glyph with the tip of her finger. The metal beneath was hot, nearly burning, but she did not withdraw. Instead, she pressed harder, feeling the glyph sear itself deeper, as if the mirror hungered for her presence.

A bead of sweat rolled from her temple and landed square on the pendant. Instantly, the glyph flared a sharp blue, subsiding into a faint, glassy gleam. The heat faded, replaced by a tingling cold that traveled up the mirror witch's hand and left her with gooseflesh. She drew her arm back, savoring the sensation of return—of boundaries reasserting themselves after the spell's brief, silent demand.

She lifted the pendant, now cool and heavy with its new burden, and turned it over in her palm. The glyph was nearly invisible on the silver, discernible only at a slant. But the sorceress could feel its presence like a slight current in the Archive's air. She tested the edges, found them as smooth as before, and nodded once to herself. A job completed, a risk well-chosen.

The last step was sealing. She produced a sliver of wax from her pouch and rubbed it along the glyph's lines, pressing the pendant into a folded scrap of cloth. As she did, she allowed herself to exhale into the stifling stillness of the room. Her work had left her skin slick and her mouth dry, but the rush of accomplishment obliterated any discomfort.

She packed away her tools in reverse order, each with the reverence of a surgeon concluding an operation, then pocketed the pendant. Before leaving the workbench, the ritualist ran her fingers over the scarred wood, feeling for any stray dust or evidence of her intrusion. Nothing. She was a ghost in the Archive, leaving no sign of the crime but the altered destiny now locked in silver.

On her way out, she paused once at the threshold. The lamp above her guttered, casting a brief shadow elongating her form along the aisle's floor. In that moment, the watcher imagined herself as the glyph: a line bent toward purpose, indifferent to consequence. She smiled, small and private, then vanished into the darkness. The pendant pulsing cold against her chest.

#

Back in the main corridors of the Archive, the veiled adept slowed her pace, acclimating to the ordinary rhythm of scholarly labor. The lamps burned brighter here, their light diffused by floating motes of parchment dust and the faintest scent of lamp oil. The pendulum clocks lining the walls ticked with absolute confidence, as if time itself dared not falter in this sanctum of perpetual record-keeping. She folded her hands into her sleeves and adopted a posture of bored diligence, the very image of an overqualified, underappreciated researcher fetching another armload of texts for review.

But beneath the borrowed calm, the pendant's chill anchored her attention. It throbbed against her breastbone, mimicked her heart, or perhaps drove it. She couldn't be sure which. The forbidden glyph, newly etched, called out for use like a muscle flexed to trembling and then denied release. The mirror witch let the anticipation build, savoring the moment before unleashing the thing she had wrought.

She paused at the foot of the stairs leading to the upper stacks. The next phase would be public, or as public as the Archive ever became. A dozen or more junior Keepers haunted these halls, their heads bent low over tablets and code-books, their minds shielded by the monotony of daily duty. Yet each one was a potential spy, consciously or not. The wards in this section—unlike those in the deeper alcoves—had been set not to maim or erase, but to alert the Watch if anything even slightly subversive occurred. They sniffed for forbidden magic with the delicacy of a wine-taster, ready to trumpet any deviation up the chain of command.

The mirror-worker extracted a copy of *Cyclical Disasters in Post-Mantle Aurelia* from a nearby cart and tucked it under her arm. She climbed the stairs two at a time, never glancing back, trusting that her shadow would not betray her. At the landing, she drifted left into a reading gallery lined with leaded-glass windows. It was empty, save for a lone scribe hunching over a sand table and an elderly man with a cluster of memory runes stuck to his scalp like barnacles.

Perfect.

She ducked into a niche near the far end, set the book on the desk, and withdrew the pendant. For a moment, she simply regarded it, letting her gaze fall through the surface rather than at it, as if the mirror might reveal her hidden intentions. Then she pressed the pad of her thumb to the glyph and whispered the activation phrase: "Obscure, revise, conceal." The words left her lips as vapor, instantly gone, but she felt them imprint onto the air with a heaviness that bordered on intoxication.

The effect was immediate and profound. The pendant's glyph inhaled the light, drawing every shadow in the room toward itself with a subtle, hungry gravity. The mirror witch's breath fogged and then vanished. The world seemed to hesitate, as if every observer—present or remote, mortal or divine—had been forced to blink at once. When perception resumed, it was with the altered lens of the spell in place: the world saw what the mirror witch wanted it to see.

She closed her eyes, testing the parameters. The illusion was robust—her own heartbeat muffled, the sense of exposure replaced by a cocoon of plausible banality. To anyone watching, she had become the model of transparent academic pursuit: the practitioner, loyal daughter, overworked and underloved, scouring the stacks for insight into the ancient cycles of disaster, just as she had been doing for months.

To test the glamour, she reached out with her awareness, probing the wards that patrolled the upper galleries. She felt their presence like a chill wind moving through water, searching for anomalies. The Mirrathar mind was adept at projecting decoys, but never before had the mirror mage felt the wards recoil from her so utterly—once probing her surface thoughts, they now slid off as if her mind were lacquered in institutional correctness.

She rose and drifted past the reading scribe, her presence failing to register even as a flicker of curiosity. At the exit, the memory-rune man looked up, his eyes glazed with the residue of a hundred borrowed recollections. He blinked, registered nothing unusual, and returned to his business. The illusionist moved through the checkpoint—one of the Archive's more paranoid additions. Without incident, the watcher rune mounted above the door failed to so much as shudder.

It worked. It *truly* worked.

A flush of victory threatened to break her composure. She reined herself in with a Mirrathar exercise; she pinched her own wrist, feeling the sharpness of bone and tendon beneath the skin, then exhaled until the urge to laugh or shout had passed. She forced herself to walk a lap of the upper stacks, feigning a search for another volume, before circling back to the gallery and seating herself at the desk.

This was the ultimate test, and it required an audience—though not one of flesh and blood.

Mounted at the far end of the gallery was a scrying panel, an opalescent sheet fixed to the wall at an angle favored by the Archive's most zealous auditors. To the untrained eye, it resembled a decorative fixture. To the mage, it was a two-way mirror, a surveillance device keyed to the oversight of both mortal and minor divine authorities.

She approached, heart racing with anticipation. The pendant's cold seemed to thicken as she stood before the scrying panel, the glyph warming in her palm until it nearly hummed. The sorceress leaned forward, her reflection forming in the glass. She saw herself, hair neatly braided, posture just a shade slumped, eyes alive with the dull fatigue of too many all-nighters.

It was perfect. Not a trace of the forbidden. Not a flicker of the true glyph. She grinned at her own image, then deliberately masked her face into a bookish humility that so often disarmed her rivals in the Council chambers. Even the Watch, even the lesser gods—always hunting for the taint of forbidden will—would see only what she chose to show.

The mirror witch lingered, letting herself bask in the spectacle of her own triumph. She pressed the pendant to her chest, letting the chill bite through her blouse, and allowed herself the smallest, subtlest bow. In the privacy of her own mind, she repeated the justification that had carried her through every risk and every betrayal: "For Aurelia's freedom." The words sounded noble in the scrying panel's silence, and she wore them like a badge.

But beneath that, beneath the layers of rationalization, another voice whispered in a register only she could hear.

Only I will do what is necessary. Only I will outwit the gods at their own game.

She left the gallery with unhurried steps, the pendant pulsing in time with her heart—a second heartbeat, secret and untouchable. The wards ignored her, the Watchers glazed over, the Archive itself seemed to relax around her presence, oblivious to the magnitude of her transgression. The ritualist could not help but smile again, wider this time. Her mask slipped for a moment as she realized the full extent of her camouflage.

The Archive's halls led her outward, back toward the world she had always intended to break and remake. The pendant, cool and unassuming, settled into place beneath her collar, its hunger sated for now but never fully dormant.

As she passed through the last set of doors, the watcher glanced at the sky beyond the Archive's dome. For a second—no more—the evening's clouds seemed to part, letting through a blue shaft of light. She considered this, measured it against her plans, and concluded that even the Loom would have to wait and see.

Her smile faded as she stepped into the new day, the illusion flawless, the secret intact.

CHAPTER TWO

Emberfield

Chapter Two

Emberfield – Twilight (*Umbersday, 5 Dawnskein, TT 1215*)

Ash fell in slow, deliberate flakes, more like snow than ruin, but it clung to Edrin Ashwood's skin with a tenderness that only the dead could grant. What remained of Emberfield stretched around him: cracked foundations, roof beams gnawed to brittle black by flame, the occasional tooth of iron jutting from the clay where a chimney had tried and failed to outlast the heat. The stench was neither smoke nor wood, but a sharper, almost metallic tang—the signature of a Loom-surge, when magic ran across the world so hot it warped the memory of what should have been fire.

Edrin stood in the midst of it all, his body carved by work and recent loss. He wore the destruction like a shroud; his clothing, once the reds and blacks of the Ashwood clan, was now indistinguishable from soot. A tall, well-muscled man, though stooped over, hunched against the grief and pain, his bared arms showing signs of deep burns from the surges when the glyph markers had started to scorch. Edrin had always worn his raven hair long, but tied back away from his smith work. Though he tried many times, he couldn't grow his facial hair like his father could, so the rangy stubble added to the appearance of more soot on the lower half of his face. His silver-steel eyes pinned everything he saw. Edrin had an appearance that made him look much older than his nearly twice-twelve threadturns. He gripped his father's hammer in his right hand, the haft sanded smooth by years of honest work, the head a dense mass of iron flecked with stars where the heat had pocked it. It was, as far as Edrin could tell, the only thing, excepting him, in the town that had survived intact.

He moved through the ruins of the forge as if he expected the anvil to get up and leave at any moment. Each step raised puffs of grey, whirling and settling, and marking his path in a landscape of the newly erased. He tested the ground ahead with the hammer's business end, nudging aside lengths of collapsed beam, sometimes pausing to stare at the striations where the wood had burned with a heat cold enough to preserve even the shape of the bark.

The anvil was there, but not as it had been. It had split along its base, a hairline fissure running up through the horn and into the face, stopping only a finger's width from the top. Edrin rested the hammer across the fracture and pressed down, as if the weight might force the halves together. Instead, the anvil groaned—a sound so close to pain that he wondered if the thing had grown a voice to mourn its own failure.

He released the pressure and let the hammer fall back into his palm. "Good try, old friend," he said to the anvil, his breath steaming in the cold that followed the surge. His voice barely carried beyond the forge's perimeter; the hush of aftermath was as complete as a prayer.

He let his eyes drift over the rest of the site. The roof was a memory. The stone walls, once mortared with pride and family debt, softened at the edges, the lime and clay slumping like the faces of the villagers who used to gather here at dusk. Only the front step remained unblemished: a slab of basalt with the family sigil chiseled deep enough that even the Loom's anger couldn't quite erase it. The hammer in his hand had made those marks, years ago, under his father's supervision.

Edrin remembered the way his father's hands guided his, shaping the symbol with slow, deliberate blows, never hurrying the job even when customers hammered on the door for new sickles or axle pins. "You never rush a mark," the old man would say. "You only get one chance to make it honest."

What was left of his father lay beneath that slab, along with the rest of the Ashwood line—bones, perhaps, if anything survived, but more likely a confusion of fused minerals and fine ash. Edrin found he had no tears for any of it. He had worked through all of that in the hours after the surge, digging for survivors, ignoring the pain in his hands as he sifted through hot rubble, calling out names he'd sworn he'd never say again.

No one answered. Not even a god.

He was the only Ashwood left in Emberfield; his distant relatives bearing the Ashwood name were far away and might never hear the call to return to Emberfield, and the hammer was the only inheritance the Loom had not managed to twist or incinerate.

He took a lap of the forge's perimeter, each step measured with the rhythm of his father's old walking song in his head. Left, right, tap with the heel, pause to test a remnant with the hammer, and on. He found a patch of stubborn flame still eating at the remains of a workbench, the fire almost blue-white and silent as it consumed a thick stack of payment ledgers. He watched it for a moment, letting the heat gather on his cheek, then stamped it out with his boot, grinding the embers until not even a glimmer remained. He would not let even a spark escape. He would not allow the Loom, or its servants, any satisfaction.

"Never again," he said, voice louder now, echoing off the stone like a challenge. "You hear me? I'll not bow to the Loom, nor to any god who loves it more than they love the folk down here."

The wind picked up, as if in answer, and sent a fresh drift of ash spiraling around his boots. Edrin ignored it. He moved back to the anvil, brushed the surface clean with his palm, and set the hammer down with the care of someone laying a child to bed. He stood there, head bowed, eyes closed. If there had ever been a right way to grieve, it wasn't written anywhere he could read.

When he finally straightened, the world had grown darker. The sun, or what passed for it after the Mantle Wars, was a dull red smear behind clouds the color of old blood. Edrin hefted the hammer and gave the anvil a single, clean strike. The sound was wrong, off-key and hollow, but it resonated through the village ruins, an announcement of a farewell.

He shouldered the hammer and started back toward what used to be the center of town. The main road was a scar of black glass, still warm in places, the occasional footprint already scuffed away by the dry wind. He passed houses that had once seemed invincible in their sameness, now reduced to low, ugly mounds. Here and there, charred bits of signboard or fence post stuck out at odd angles, pointing nowhere in particular.

At the crossroads, Edrin paused. The town's marker post had survived—miracle or oversight, he didn't care. The village name, once proud and painted in fire-resistant ink, now barely legible:

Emberfield. He spat into the dust and turned left, taking the shortcut through the abandoned market square.

There, amid the toppled stalls and scattered fruit husks, he found something new. Tracks—light, but real. Not the heavy tread of a villager, nor the delicate print of a scavenger or stray animal. These tracks were deliberate, almost ritual in their evenness, leading directly to the one building left standing in the square: the old mill.

Edrin gripped the hammer, feeling the sweat rise on his palms despite the cold. He followed the prints, each step landing exactly in the hollow left by the previous visitor. The mill door hung on a single hinge, its boards scorched but intact. He pushed it open with the hammer head, ready for anything.

Inside, the air was different; moist, almost sweet, the smell of milled grain somehow persistent even after the world had burned. At the far end of the room, crouched by the rusted mechanism of the waterwheel, was a figure: large, hunched, hooded, hands busy with something Edrin could not see.

He advanced, slow but unafraid. If it were a looter, they'd be more trouble for the hammer than the other way around. If it were one of the Loom's priests, better to settle their account here and now.

But the figure looked up and rose from his haunches as Edrin approached. The smith saw not the blank mask of a priest, but the scaled face of a Lizardkin, a rare species anywhere near Emberfield, though occasional soldiers from the Sytheri homelands had been stationed around the village in the old days. The stranger looked young, but the eyes were clear, brightly purple with vertical pupils, though they were tough to see completely through some oddly glowing glasses. He was fixed on his surroundings with a focus that made even the last survivor of Emberfield feel transparent. As Edrin closed, the silver-tailed Lizardkin began to speak. "You are late," said the Lizardkin, voice reedy but not unkind. "The surge registered at dawn. I expected a response team, not one man."

Edrin rested the hammer on his shoulder and glared. "You expected wrong. Emberfield's dead. I'm the only one left to answer."

The scholar nodded, as if this was neither a surprise nor a disappointment. "The Loom will be unhappy. This event was... significant. The pattern is altered. The Archives will need a full accounting."

Edrin's jaw clenched. "Let the Loom choke on its own pattern. It's not getting anything from me."

The Lizardkin looked down, fiddling with a rolled parchment. "I suppose not," he said. "But the world needs what you know. And you need what I can offer."

Edrin studied him, weighing options. He had nothing left to lose—no family, no home, no reason to care. But the hammer in his hand felt heavier than it ever had, and the words he'd spoken in the ruins lingered in the air like a promise.

"What's your offer?" he asked.

The scholar unrolled the parchment, exposing a map marked with glyphs and strange, curling annotations. "There are others like you," he said, voice barely above a whisper. "Others who survived what should have killed them. The Loom wants to erase them, but I want to study them. I want to know why the world resists."

Edrin let the idea settle, but not for long. "And if I say no?"

The Silvertail shrugged, as if refusing gravity. "Then I follow you, and the Loom follows me, and eventually, we find out the answer together, in whatever way the world prefers."

Edrin laughed, the sound harsh but genuine. He liked this Lizardkin already—there was no lie in him, no pretense. He was as broken as Edrin, and maybe just as angry.

"All right," Edrin said, shouldering the hammer. "Lead the way, Scholar. But if the Loom wants a fight, it's going to get one."

The Lizardkin stood, joints creaking. "Bran," he said. "My name is Bran Silvertail. You'll want to remember it." He flicked his forked tongue through the air briefly, "The world is about to make us both famous, or infamous, or dead."

Edrin nodded. "Edrin Ashwood. Last of my line. And I don't care much for fame. I am already dead." He indicated the dead husk of the town around him.

They stepped out into the night together, the forge ruins behind them, the Loom's war ahead.

As they walked, Edrin, for the last time, looked back at the place where his home had been. He thought of the hammer, of the anvil, of the glyph-scars on his body. He let the memories settle, then pushed them away. There was no room for them anymore—not if he was going to do what he had promised.

"Never again," he whispered, quieter this time, but just as final. "I will not bow."

He ground the words into the dirt with his heel, and the world—silent, ruined, and watchful—took note.

#

The hour before dawn found Edrin scavenging the last valuables from the forge, a task part necessity, part ritual, part raw defiance. He kept the hammer at his side, but his hands wandered the blackened rubble for anything else the surge had spared: a length of tempered steel, barely warped; a bundle of wires for snaring; a handful of coal, still hard and unburned at the core. He collected these into a battered knapsack, not because he needed them, but because leaving anything useful to rot felt like a second surrender.

He was bending to pry a gear loose from the kiln when the scrape of footfalls on glass set his nerves alight. Edrin whirled, hammer ready, only to find Bran Silvertail standing not a dozen paces behind. The lizardfolk's skin was a cool, liquid silver in the morning's first ghostlight, and his gaze, unblinking, seemed to take in every detail at once.

"You follow quietly," Edrin said, not trying to hide his annoyance. "You spying, or just waiting to see if I'd get myself killed?"

Bran tipped his head, an expression that might have been amusement in another species. "Neither. I was examining the perimeter for residuals. Your movements are… efficient, if a bit loud."

Edrin scoffed, dusting his palms on the thighs of his ruined trousers. "Loud's better than dead. What's it matter, anyway? Town's empty. Even the crows skipped out after the surge."

Bran made no reply, but knelt in the debris, producing from his pack a leather tube bristling with scrolls, slates, and measuring instruments. He began to unroll a parchment, weighting the corners with pebbles. The chart's surface was a chaos of lines and glyphs, but at its center lay a spiral—a perfect, endless coil—etched in black.

"You keep doing that," Edrin said, watching Bran align the spiral with the remains of the kiln. "Drawing circles. You always that obsessed?"

Bran's tongue flicked, tasting the air. "They are not circles," he corrected. "They are patterns. And this one matches the site geometry exactly." With a careful hand, Bran drew a polished bone stylus from a case and traced points on the parchment, each one corresponding to a landmark in the ruins.

Edrin grunted, but moved closer despite himself. "What's it mean?"

Bran's eyes, cold and lambent, didn't leave the map. "It means this was not a random surge. The energy followed a deliberate path—one that's been replicated elsewhere, and always with the same aftermath." The stylus darted, annotating dates, signatures, and odd, technical glyphs.

"Let me guess," Edrin said, "the Loom did it. Some pattern in the big weave, right?"

Bran looked up, face blank. "Not the Loom itself. The Loom is the substrate. What acts on it is… more complex. There are footprints in the energy. This one—" he jabbed at the spiral, "—matches the old records. Matches the prophecies from

Threadmere. Matches the dreams the seers have been having since the Mantle Wars."

Edrin snorted. "Dreams. Prophecies. Sounds like a load of—"

"The Silvertail archives do not lie," Bran said, voice flinty. "Neither do the residuals." He produced a small device—a prism wrapped in wire and crystal—and held it above the ground. At first, nothing happened, but as Bran muttered a sequence of numbers, the device glowed faintly, then cast a pattern of intersecting lines on the dust. The lines matched the spiral, matched the glyphs, and matched the burn marks left in the earth.

Edrin's skepticism melted a little. "All right," he allowed, "that's clever. What's it supposed to prove?"

Bran coiled his stylus between his claws, clearly savoring the moment. "It proves there is a hand behind these surges. A signature. And it is not divine—not even the gods work in motifs this precise." He began to pack up the device, rolling the parchment with obsessive care. "This site is the fourth I have charted. Each one stronger than the last. Each one closer together in time. And in every case—" he paused, making certain Edrin was listening, "—there is a survivor. Always only one."

Edrin glanced down at his hammer, then at his arms, where the scars from the burns seemed to glow under a thin film of sweat. "So what, I'm cursed? I already knew that! I just recovered the ash, not the bones, just the ash of my entire family."

"Not cursed," Bran said, "marked. Like a variable in an equation." He looked at Edrin with a strange intensity. "Have you never wondered why you survived? Why this hammer survived? Why did the Loom not finish the job?"

Edrin found he had no answer for that. He wanted to say luck, or fate, or some other word that didn't feel so loaded, but nothing came. But he also knew that he didn't feel special, just lonely and furious.

Bran set the scroll back in its tube, stood, and brushed off his knees. "We should travel to Threadmere," he said. "The scholars there have deeper records. And the Veilwood seers might be able to confirm if this—" he gestured at the ruins, "—is an isolated phenomenon or the start of a new cycle."

Edrin bristled at the suggestion. "And why would I help you? I've got no stake in prophecy or whatever the gods are gambling with down here. I am going to wait for others and help build this place back up. I have distant cousins, other families who can help." His voice trailed off as he realized that he didn't believe in what he was saying.

Bran smiled—a taut, almost painful thing. "You said it yourself: you will not bow to gods or Loom again. I am offering you a chance to prove it. Not with words, but with action."

Edrin gripped the hammer tighter, feeling the callus shift beneath his thumb. There was a part of him that wanted to walk away, to vanish into the far hills and let the world chase its own tail. But another part—older, more stubborn—wanted answers. Wanted to hit back at whatever force had twisted his life until nothing remained but the grip of a tool.

He scuffed at the spiral with his boot, but the lines would not erase. "Veilwood first, then," he muttered, "then on to Threadmere after. "

"Agreed," Bran said, and Edrin could not help but feel the satisfaction rolling off him like a heat shimmer. "We will move after dawn when there is good light."

Edrin grunted. "Fine. But if I find out you're using me for some lizardfolk ritual, I'll smash your device to bits and dump the pieces in the river."

Bran bowed, but there was no sarcasm in it. "Understood. Now, may I suggest you rest a bit? The sunrise is a short time from now, and the journey will be long, and the Loom is not done with us yet."

Edrin watched as Bran padded away, every step measured, his tail making faint, perfect trails in the ash. Edrin found a patch of ground that wasn't still burning, lowered himself onto it, and tried to ignore the fact that the spiral Bran had charted was now burned into the inside of his eyelids.

He stared at the sky, at the not-quite-stars bleeding through the haze, and wondered if the Loom ever slept. He doubted it. He doubted he would, either.

But as he closed his eyes, the only thing he saw was the hammer, glowing steady and true in the darkness, a promise of resistance that no prophecy could break.

#

Edrin woke to find Bran already awake and working. The lizardman had laid out a careful spread of supplies on a flat stone: the rolled maps, the prism, a stack of dried food, and two canteens that had been rinsed and refilled from a spring Edrin didn't remember existing.

Bran was triple-checking his inventory, methodical as a crow picking at a bone. Edrin watched for a time, letting the morning seep into his bones, then rose and slung his knapsack over one shoulder, the hammer tapping against his thigh.

They didn't speak at first; there was nothing more to say that hadn't already been burned into the night. Instead, they walked the perimeter of Emberfield one last time, Edrin scanning the ruins for anything he'd missed, Bran studying the patterns in the scorched earth as if reading tea leaves from the ashes.

At the village boundary, where the marker post still stood, Edrin paused. The sense of departure was heavier than he'd expected. He stared at the warped letters of the town name, at the blackened fields beyond, at the silent nothing where his mother's garden had been. He squeezed the hammer, feeling the bite of the wood grain against his palm.

Bran, sensing the moment, halted beside him. "You don't have to come," he said, with the dispassion of someone who'd rehearsed the argument and lost it long ago. "But if you want answers, or revenge, or even just a reason to keep moving, you won't find it here."

Edrin grunted, but there was less fight in it than before. "No. There's nothing left here except memories. And I'd rather not have the Loom decide how I remember."

They set out for the road. After a time, Edrin broke the silence. "What exactly does this 'Threadbreaker' prophecy say? You keep dropping hints, but never the whole thing."

Bran hesitated—rare for him. "It's not a single prophecy," he said. "It's a motif, threaded through a dozen cycles of vision and history. Theoretically, it happened once, but the records are either hidden or forbidden by the Mantle Council and the archivists. But the heart of it is simple: that one day, a mortal will break fate's threads. Not just bend them, or dodge them, or get away with something clever. Break them, and rewrite the whole pattern."

Edrin turned this over in his mind. "And the Loom's scared of that?"

Bran nodded, the scales at his throat rippling in the light. "It is why the surges happen. Why does it try so hard to snuff out the variables before they reach critical mass?" He gave Edrin a sidelong look, not unkind. "I postulate that it's why you survived."

Edrin let the words settle. He looked at the hammer, at the glyph burns on his arms, at the burned horizon behind them. "So what, you think I'm this 'Threadbreaker'? That's rich. I'm just a smith who happened to be too stubborn to die."

Bran's tongue flicked, a smile hiding in the gesture. "That's how all these things start. Someone gets too stubborn to die, and then the world changes around them." He shouldered his own pack, the bone stylus tucked behind one ear, his glyphic glasses slightly askew on his face. "Besides, if it's not you, it will be someone else. The Loom doesn't get to have everything its way. That's what the pattern says. For all we know, we are doing what the Loom wants. Or we are doing exactly what it doesn't want. Fun, eh?"

They walked in silence for a while longer, the ruins of Emberfield shrinking behind them. As the sun crested the low ridge to the east, Edrin stopped and turned to Bran.

"If we're going to do this," he said, "then we do it as equals. No more of this 'subject and scholar' nonsense. You want to study the Loom's handiwork through me; you show me everything. No secrets."

Bran's eyes glinted. "Agreed. Equal partners, for as long as the pattern holds."

Edrin spat into his palm, then offered his hand. Bran looked at it for a second, then copied the gesture, his own saliva stringy and iridescent. They shook—a grip hard enough to crack bones, if either of them had been inclined to let go.

The pact made, they turned toward the distant spires of Threadmere, the road ahead still pocked with the scars of war and the leftover magic of a world trying to mend itself. Edrin walked with the hammer at his hip—a weight, a comfort, a reminder, a promise.

As they left the boundary of Emberfield, Edrin looked back one last time. The village was already losing its shape in the morning haze, but he could still make out the anvil's silhouette, standing alone in the rubble. He raised the hammer in salute, then turned away, refusing to let the past define his stride.

The road to Threadmere was long, and the future uncertain, but Edrin walked it anyway because the alternative was to stand still and let the Loom finish what it had started.

And for the first time since the surge, he almost looked forward to what came next.

CHAPTER THREE

Veilwood Hollow

Chapter Three

Veilwood Hollow – Morning (Umbersday, 1 Mistbloom, TT 1215)

The long journey to Veilwood Hollow gnawed at Edrin's patience. The first day was all packed loam and cold wind, the kind that whistled between the ruins and froze sweat to his spine; the second was a relentless trek through the backcountry, where the old forests were tattered and ripped, remembering the Mantle Wars better than any living soul. And it got less and less familiar the days and weeks on the road. Bran led with a scholar's rigor, never pausing longer than it took to annotate a tangle of glyphs scratched into a trail marker or calibrate his weathered chronometer. If he was tired, he never showed it, though Edrin sometimes caught him humming in quiet sympathy with the woods, as if the old songs might appease whatever still prowled these hollows.

The land changed as they closed on the Veilwood. The air grew heavier, saturated with the green hush of sap and the ghost of centuries' prayers. The trees, mostly tallow-elm and sorrowbirch, arched in patterns that looked deliberate—no two branches the same, yet every one angled to admit a certain slant of sun, as though the forest itself had been trained to think in geometry. Moss spread in thick mats across the ground, disguising the lay of the land and muffling their approach until even Edrin, so attuned to the clangor of the forge, felt himself begin to vanish into the quiet.

They finally arrived on the morning of Umbersday, 1 Mistbloom, the precise day the Lierabraid festival began its first spiral for the Sunweft season. The approach was marked by increasing signs of human effort: branch-woven effigies perched atop boulders, flags of woven bark strung across gaps in the path, stones set in runic spirals that pulsed with embedded glowstones. At each new pattern, Bran's eyes shone with something close to hunger. Edrin's, by contrast, narrowed. His mistrust of organized magic was almost instinctive now, after all the suffering in Emberfield. As the path opened into the ritual grove, he tensed involuntarily, as if the trees themselves might sprout hands to drag him into the ritual circle.

The festival had already bloomed into a full ritual. The seers—at least thirty of them, by Bran's count—circled the central standing stones in concentric arcs, their robes heavy with the dust of old pigment. Each robe was unique: some were indigo edged with gold thread, others were ashen white. The robes themselves moved with a collective life, rippling in synchrony even when their wearers stood still. Edrin searched for the source of the choreography—some signal, some invisible conductor—but only found the uncanny group so attuned it had become an instrument.

Braids of colored thread, thick as a blacksmith's pinkie, hung from the seers' waists and wrists, each strand alive with a soft, internal glow. As the seers walked, they braided and unbraided the threads, at times joining a neighbor to weave a more complex strand. At each intersection, the colors brightened, momentarily spiking in luminescence before subsiding into a steady gleam. The effect was something between magic and animal behavior, a hive mind rendered visible in silk and light.

Bran hovered at the edge of the crowd, chin tucked low, every motion of his clawed hands precise as a surgeon's. "They time the braids with the lunar clock," he whispered, mostly to himself but loud enough for Edrin to hear. "Every twenty-eight breaths, a new thread is joined. Observe: the synchronization is not perfect, but the error margin narrows with every cycle."

Edrin watched a young acolyte—barely older than a child—fumble a loop and redden as her superior gently corrected her grip. He grunted, unimpressed. "Looks like a lot of wasted labor. You ever try mending a real net with a crowd that size? The whole thing would snarl in a heartbeat."

Bran's tongue flicked out, catching a mote of incense in the air. "Not if the net was the point, rather than the product. They're not weaving an object; they're weaving the illusion of order. The act itself is the sigil." He made a note on his slate, then pocketed the device with a flick of his wrist. "It's rather elegant, in a way."

Edrin snorted but held his tongue. He shifted the weight of his hammer at his belt—a comforting motion, though he knew full well that steel was the last thing that mattered in a place like this.

Incense smoke curled around their ankles, rising from a hundred shallow braziers spaced at precise intervals along the mossy stones. The smoke was gray, blue, and gold, but the scents never quite aligned with the colors. Edrin's nose picked up undertones of burnt cinnamon and sharp nettle, then a sweetness that turned his stomach, reminiscent of the offerings left at Emberfield's old shrines. At the heart of it all was a current of voices: not a song, but an underchant, the syllables blurring and melting until they became the resonance of the grove itself. Now and then, the chant would crest in a single, unified syllable, "Aelion," "Tirra," "Veilos". Then collapse back into babble.

Moss and lichen carpeted the ritual ground, turning every stone into a muted green. The standing stones themselves were ancient, each at least twice Bran's height, their surfaces scored with glyphs that shimmered faintly beneath the touch of the seers. It was difficult, even for someone as uninitiated as Edrin, to ignore that the stones were aware, watchful, or perhaps waiting.

A horn sounded from the far end of the grove, a single low note that sent a ripple through the gathering. The seers moved inwards, the outer rings folding toward the center like petals closing at dusk. Edrin's shoulders rose a fraction higher.

Bran, unfazed, produced his chronometer and timed the contraction of the crowd. "Predictable," he said. "Every generation, the ritual accelerates. Used to be, the entire day would pass before the closing. Now, three hours. Four, at most." He scribbled in his notebook. "Fear of time running out, perhaps?"

A pair of children in pale robes skittered past Edrin, their arms full of loosely braided thread. The threads trailed behind, luminous and snakelike, lighting the moss with splotches of color that reminded Edrin uncomfortably of cauterized wounds. He watched them vanish behind one of the larger stones, where a handful of elders waited, hands outstretched to accept the offerings.

Edrin looked up at a woman wearing a robe of deep forest green, her eyes marked by a series of silver tattoos radiating outward from the pupils. She was older than most present, her hair twisted into a crown of twin braids looping behind each ear.

She examined Edrin in silence, as though cataloguing each of his scars and drawing conclusions from the way he held his arms.

"You are not of the Hollow," she said, her voice gentle but edged. "You walk with the mark of ashes."

Edrin's grip shifted to the hammer, but he made no move to draw it. "Passing through. The Hollow isn't closed, is it?"

The woman smiled just enough to show that she'd heard every lie ever told on these grounds. "Not to the desperate, nor to the lost. But you should not approach the stones unless you are ready to leave a part of yourself behind." She glanced down at his hands. "Or perhaps, to reclaim one."

Bran stepped between them with a deference that bordered on comic. "We are guests only," he said, bowing his snout. "My companion bears the residual of a recent surge. We seek an audience with your seers, in the hope that they might—"

"—see what the gods refuse to show," the woman finished. "Yes, I know." She leaned in, lowering her voice. "You are not the first to come this cycle. But you may be the last."

She pointed to the moss ring closest to the standing stones. "Sit. Wait for the Lierabraid to complete. Then, if the Hollow accepts you, you will be called."

Edrin and Bran took their places on the moss—Edrin, cross-legged, tense, as if the ground could give way at any second. The ritual continued to tighten around the central stones, the braids of light now so dense they formed a living net above the heads of the innermost ring. The chant grew louder, the syllables resolving into a pattern that even Edrin could almost recognize. It was a language older than memory, but the intent was clear:

containment, warding, hope in the face of something vast and dangerous.

As the braids joined in the center, a sudden burst of light shot up from the stone, refracting through the woven threads and painting the canopy with prismatic arcs. The effect was dizzying, almost hallucinatory, and Edrin felt his vision double for an instant. The world tilted sideways, then righted itself with a jolt of clarity.

The festival reached its apex in a thunder of color and song, then just as suddenly collapsed into silence. The seers stepped back, releasing their braids to the earth, where they smoldered in silence before dimming to nothing. Only the echo of the chant remained.

Edrin let out a breath he hadn't realized he'd been holding. Bran looked to him, an eyebrow ridge raised, but said nothing. Across the circle, the woman in green waited, her eyes fixed on Edrin with patience.

The Lierabraid was complete. Whatever happened now would be between Edrin, the Loom, and whatever truths the Hollow cared to share.

#

When the last echo of the Lierabraid died, the seers carefully retreated from the circle and disappeared. The crowd scattered into knots of private discussion, each group quietly dissecting the ritual's omens, tallying the tiny failures and proud successes. Yet one figure stood motionless among the shifting bodies. A young, pretty woman whose robe of cloud-grey was almost lost against the pale lichen coating her seat. She seemed built of soft intent and transparency, the way a stray sunbeam is barely there, except for the clarity with which it strikes the eye.

Edrin would not have noticed her at first, but as the color and noise drained from the grove, her steadiness caught his attention. She sat alone, close enough to the standing stones to draw rebuke from the older seers, but nobody challenged her. A hood framed her face, its fabric so fine it hinted rather than concealed. She pushed her hood back, revealing moonlight white hair with a shimmer of midnight blue, braided so intricately that it seemed more sculpture than ornament. Her features were gentle, but a tracery of thread-thin glyphs mapped her temples and the hollows beneath each eye, like rivers charted in the fragile skin of a hand. Her hands themselves were remarkable: large-knuckled and square, the nails bitten short, the left index permanently bent at a slight angle, as if it healed poorly after being broken.

She was working a length of thread between her fingers, silver and blue and almost too bright for the shade. The braid took shape with swift, repetitive motions, each twist perfectly timed, the tension and slack measured by a mind elsewhere. Edrin watched her for a full minute, uncertain whether she was in a trance or simply ignoring the world.

Bran noticed his gaze. "That is Lysera Vayesh," he murmured, careful not to be overheard. "She is well-known. Some say she was born on a day with two full moons and that the Hollow's glyph-mother touched her at birth."

Edrin glanced sidelong. "I suppose that means something around here?"

Bran shrugged, his crest rising and falling. "Only that she Sees more than most, and pays the price."

As if on cue, Lysera's hands stilled. Her head tilted, as though she was listening to something moving inside the stone. The moment stretched until it threatened to snap, then her body jerked in a tiny, involuntary shudder. The silver-blue braid fell from her hand, forgotten. Her eyes unfocused, pupils dilating, irises flickering with a pale, unnatural light matching the pattern of the threads. The skin around her temples prickled, the scars standing out in sudden relief; her breath fogged in the cool air, though she sat beneath a shaft of honest sun.

Lysera's vision arrived as a rush of competing senses—smell, sound, touch, all at once. She saw the ritual spiral still twisting through the air, but instead of fraying as it should, the energy collapsed inward, sucked toward a single, predatory void. The faces around her melted to shadow, except for one: a featureless mask looming above the world, its eyes stitched shut by thick black thread. Invisible hands plucked at the spiral, unraveling it with careful, hungry precision. In the far periphery of vision, Lysera saw a thousand identical strings, each taut with a future she could not yet name, all converging on a single instant of collapse.

She blinked, and the world shuddered back into place. The smell of burnt incense returned, the sound of seers chattering, the cold ache in her knees from sitting too long. Her braid lay on the moss, a limp and ordinary thing now. But the knowledge of what she had Seen pressed at her ribs, sharp and insistent.

Lysera looked directly at Edrin, as if she'd always known where he would be. Her voice was low and measured, not the ethereal whisper he expected but something solid and specific.

"You should not linger here, Edrin Ashwood. The world will need you soon in many other places, and the longer you delay, the tighter the threads will bind."

He bristled at the prophecy, shaking his head in a slow, deliberate arc. "If you're here to tell me my Loom's fate, save it. I don't bow to gods or their errand runners."

Her smile was sad and entirely without self-pity. "You misunderstand. I do not serve the Loom. I've spent my whole life resisting it. But what I saw just now—the spiral? It is not finished with you. And if it finds you alone, you will die. This time for good."

Bran interjected, "You Saw this in the ritual?"

Lysera nodded, gathering the threads from her lap and winding them around her fingers. "There is something behind the pattern now. It pulls, it shapes, it harvests the moments when things are weakest—when the cycles are ready to fracture. It wants a particular ending, and it will grind the world to powder until it gets it." She met Edrin's gaze again, unflinching. "But you, and others like you, make the Loom uncertain. You disrupt the calculation."

Edrin scowled. "So now I'm some kind of counterweight, is that it? I'm tired of being told I'm important just so you can shove me where you want."

Lysera shook her head, the motion slow and heavy. "That is the point. No one can make you do anything—not the gods, not the Loom, not even me. But if you do nothing, the world ends on its terms. If you act, maybe it doesn't."

He stared at her, anger fighting with a grudging respect. "You're not wrong. But what exactly do you think you can do? You can See, but can you change anything? Or do you just pass along the bad news to the next poor bastard in line?"

Lysera's laugh was the first bright thing Edrin had heard all day. "I can do more than See. I can walk beside you. Guide you when the world tilts sideways, and maybe keep you alive long enough to see your own ending." She leaned forward, voice dropping to a conspiratorial hush. "Or do you want to face what's coming alone?"

Bran was already nodding, the notion of traveling with a seer clearly appealing to his appetite for data. "It would help," he said, "to have someone who can anticipate the next collapse before it happens. The pattern adapts, but it cannot hide from your kind."

Lysera gathered her braid and stood, dusting moss from her knees. She was not tall, but something about the way she moved made the space around her seem to expand, as if she carried her own bubble of inevitability. She tucked the silver-blue thread into a pocket and fixed Edrin with an almost challenging gaze.

"I won't slow you down," she said. "But I need to come with you. The threads show a pattern larger than any one of us. What I've Seen must be shared with those who can act."

Edrin's jaw tightened, the muscle standing out like a scar beneath his ear. He tried to conjure an argument, something that would let him keep his quest clean and solitary. But the truth was, every disaster he'd lived through had been a team effort, whether he liked it or not. He was too old, in the way of trauma, to believe in the myth of the lone hero.

He nodded once, the motion abrupt and resigned. "Fine. But no more of this destiny business. I'm not a symbol. I'm not a 'chosen' anything. I'll smash the Loom myself before I let it use me again."

Lysera's smile softened. "Agreed."

Bran grinned. "Then we have a quorum."

The trio stood together, ringed by the moss and the old stones. Around them, the festival dissolved into ordinary time—children laughing as they chased glowing threads, elders debating the minutiae of this year's omens, the last of the incense curling through the trees like a ghost unwilling to depart.

Lysera stepped into line with Edrin and Bran. The world, for a moment, seemed to pause, as if waiting for their first step.

"Where now?" Lysera asked, her tone half-curious, half-knowing.

Edrin squared his shoulders and set his sights on the break in the trees leading to the north trail. "We go to Threadmere," he said. "If there are answers anywhere, it's in the Mantle Council's records. The council should be told, and we should have enough proof to get some research time in the Archives. And if not there, we'll keep walking until something worth fighting shows up."

He set off, Bran and Lysera falling into step beside him. Above them, the sun sifted through the boughs in slanting gold, and the last prismatic echoes of the Lierabraid faded.

They left the Hollow with three pairs of footprints in the moss, side by side, and a new pattern beginning to wind its way through the world.

#

In the oldest cellar beneath Threadmere's north quarter, behind a door that required two passwords and a drop of blood, the mirror witch kept one of her own sanctums. She called it the Mirror Room; it was as much a laboratory as a shrine. Every surface that was not glass was stone—a cold, unyielding granite veined with pale threads of fossilized stoneborn lava eels. The only source of heat was a single oil lamp, shielded in blue mica, and the only sound was the soft, relentless pulse of the mirror witch's own breath as it bounced back from a thousand reflections.

The mirrors filled the room in patterns of exacting precision. The largest—a full-length arc of polished obsidian—anchored the far wall. Smaller hexagonal mirrors tessellated the other three, their silver backs exposed only at the edges to maximize the effect. Even the ceiling bore a lattice of angled shards, set to reflect certain points and cancel others. It was a place for secrets, not for beauty.

The veiled adept stood at the nexus of the pattern, hands clasped before her, the new pendant resting at the hollow of her throat. Her hair, usually tamed into a severe knot, now fanned out in blue-violet wisps as if electrified by the room's potential. Her robe was less elaborate than the one she wore at Council, but its fabric shimmered with embedded glyphs moving only when one was not looking directly at them. Slowly, she shouldered out of the robes, which left her wearing nothing but the pendant around her neck. She was tall with many curves, though her face looked slightly hollow and pinched. Almost catlike, she studied her reflections all around her, admiring the look of the light reflecting on her and back to the mirrors.

She breathed in once, the air sharp with the tang of cold and the faintest hint of ozone from the old wards. Then she set her mind to the task.

With her left hand, she drew a circle in the air before the central mirror, a gesture so practiced it was almost lazy. The glass responded at once: its black surface brightened, then dilated, like an iris opening to admit more light. Within the growing aperture, a flicker of otherness began to stir. First, a simple shadow, then a cascade of ambiguous forms, some tall, some squat, all vibrating at a frequency that ached at the edges of the mirror-worker's vision.

She pressed her thumb to the pendant's glyph. The etched spiral, so faint it was visible only at certain angles, flared with a quick, icy blue. The effect passed into the glass, where the shadows coalesced into the approximation of faces. Each was a mask, yet behind each mask lay an intelligence old enough to make the stone shiver.

"Your time is precious," she said. The words fogged in the chill, but the voices from the mirror overlapped in reply.

The mirror witch inclined her head just enough to acknowledge the point. "I request an audience only to test the new lattice. You may probe as you wish."

The faces pressed closer to the surface, each eye a pinprick of cruel curiosity. One—a mouthless, elongated oval—spoke in a voice like shattered glass. "You call upon me to decipher your spells? Me?!?"

The mirror witch permitted herself the faintest smile. "It is of the highest necessity. You desire results, I desire latitude."

A pause, filled with static and the faint, nauseating smell of sulfur. The mirror vibrated violently, the dust lifting from the seams in the floor.

The assault, when it came, was purely psychic but so intense that it nearly dropped the practitioner to her knees. The minor gods—shadowed offshoots of the Loom, not the primary Seven but some of their enforcers—struck at every soft place in her mind. They probed the glyph's logic, each layer of misdirection and obfuscation, looking for any hint of her true plan. The mirror mage steeled herself, focused all thought on a single wordless image: the braid she had studied in the Archive, the one that had changed the rules for what a glyph could be. She held the image and nothing else.

The faces in the mirror blurred and doubled, then thinned into the outline of a single hand, palm outward, fingers spread. In the pattern of the palm, the illusionist glimpsed the very glyph she had etched, reproduced perfectly. She felt a moment of genuine panic—had she been read, revealed, ruined?

But the hand reversed itself, and the glyph on the palm became a spiral with its center turned away, closed, a sign of acquittal.

She let her breath out, then nodded. "I accept the risk."

The mirror darkened, the faces receding into memory, the surface flattening until it was again merely glass and shadow. The mage lifted her hand from the pendant, the cold in her fingers tingling all the way to the bone. She allowed herself one long exhale before straightening her posture and turning to the next task.

She approached a smaller mirror, set at eye height on a west wall, and breathed onto its surface. The fog cleared to reveal a different kind of image: three tiny figures walking a green-blanketed trail. Even at this distance, the glyph burns on Edrin's arm were unmistakable, and Bran's metallic scales caught the light as he tilted his head to address their companion. The third—a girl in grey, hair like a cut thread—walked beside them. The mirror witch watched as they moved through the world, unknowing and unafraid.

She placed two fingers on the glass, careful not to leave a print, and whispered: "See what you will. It changes nothing."

The surface rippled, as if the three had glanced up to meet her gaze, then went still.

The sorceress straightened, surveying the full hexagon of mirrors. For a moment, the silence returned—absolute, perfect, but for the echo of her own breathing. Recovering the robes and donning them, she checked her pulse, still fast, and tucked the pendant beneath her robe.

"Not bad," she said, this time to her own reflection, which looked back at her with a composure that bordered on arrogant.

She turned out the lamp, leaving the room in perfect darkness, then listened as her footsteps faded away.

When she was gone, the mirrors remained, each one holding its surface as if waiting for the next act to begin.

CHAPTER FOUR

Veilwood Hollow Leyline

Chapter Four

Veilwood Hollow Leyline – Night (Umbersday, 15 Mistbloom, TT 1215)

The leylines unfurled beneath their boots as a corridor of impossibility—no two steps ever the same, no two breaths drawn against quite the same air. Some said the Paths cheated distance, others that they cheated sanity. Bran insisted that the leylines were the only rational path to reach Threadmere from Veilwood Hollow in less than a month's march. However, Edrin doubted the words 'rational' and 'leyline' had ever consented to be in the same sentence together.

They moved in tight formation, each one bristling with their own brand of nerves: Bran in the lead, eyes darting for leyline markers and every ripple in the Paths' blue-lit geometry; Lysera with her head tilted back, as if auditioning for a choir of ghosts only she could hear; Edrin in the rear, hammer slung at his hip and eyes fixed on the places where the road looked ready to drop them into the sky.

There was no sun while walking along the leylines, only a cold, pulsating shimmer overhead—a vein of light that sometimes forked and bled new roads into the emptiness. The surface underfoot was neither earth nor stone but a half-fired ceramic—brittle and too smooth, like sun-bleached bone. When he looked sideways, the world bent at odd angles; he tried not to look sideways.

For a time, their journey was merely a lesson in patience and paranoia. Bran kept stopping to cross-reference his own notes with what the leyline itself told him. Every few hundred paces, he'd reach into his coat for a battered brass chronometer, tap it against a glyph-etched stone, and scrawl something into his field journal.

"I don't see how writing about the world helps you survive it," Edrin grumbled the third time Bran halted them. This time, however, he measured the air temperature with a strip of color-shifting paper.

Bran didn't look up. "It helps if you wish to survive more than once. Besides, we are not merely surviving; we are observing a phenomenon." The lizardfolk's tongue flicked, tasting something Edrin couldn't see. "The leylines are less stable here than expected."

Lysera smiled a soft, tired smile. "Is that your way of saying we should hurry?"

Bran's scales shimmered with a nervous metallic sheen. "The next bridge will collapse in approximately one hour," he replied. "If we reach the convergence before then, we may yet avoid—"

He stopped, tail rigid.

Lysera was already backing away from the centerline, her skin gone pale. "It's starting," she said.

Edrin felt it before he saw it: a prickling behind the ears, the sense that his bones had been swapped for someone else's. The world tilted, then cracked.

It started as a low, electric whine. The sky above the leyline split open, not with light, but with lines. Dozens, then hundreds, of phosphorescent spirals spun out from a single point of blackness like augers drilling into space. The air grew thick, charged, like the inside of a thundercloud, and every hair on Edrin's arms stood at attention. The whine grew sharper, then fractured into a scream that only the metal in his hammer seemed to answer.

Stitch-storm.

The first bolt of the storm arced across the path ahead, a whip of violet lightning that licked the ground and left a sizzling trench in the stone. Bran yanked his chronometer from his belt, pressing it flat to the nearest marker, his other hand clawing at his notes as if speed-reading would give him an edge over the disaster.

Lysera's eyes glowed, the whites gone completely, her irises reduced to twin disks of rolling silver. She pointed with a trembling finger at a break in the path's geometry—a place where the leylines buckled, folding in on themselves like origami.

"There," she said, her voice layered with the timbre of three people at once. "If we run now, we can cross before the threads overlap."

Bran didn't hesitate. "Go. Now."

They ran. The stitch-storm broke over them in waves: arcs of pure color slashed across their faces. Sometimes the path underfoot vanished, replaced by a hungry darkness. Sometimes the path doubled, and Edrin had to choose—left or right, knowing that picking wrong might drop him a hundred feet or into a year that wasn't his.

Lysera kept to the centerline, always a step ahead, her Sight burning with intensity. Edrin gritted his teeth and followed, ignoring the pain as a bolt clipped his shoulder and left a numb, icy burn in the muscle. Bran trailed just behind, eyes darting between his chronometer and Lysera, his notes now flapping behind him in a trail of loose pages.

Halfway across the shattered segment, a rift opened in the path—a physical wound in the world, wide as a river and pulsing with a sickly green light. Lysera hesitated, the glow in her eyes flickering.

"It's not supposed to be here," Bran hissed. "This wasn't in the charts. It's—"

The wind hit them then: a blast so cold and strong it flattened Edrin's chest and knocked Lysera to her knees. Edrin dropped the hammer, wrapped his arms around Lysera's waist, and hauled her up just as the rift's edge began to unspool. He half-carried, half-dragged her the last dozen paces, only to find Bran already on the other side, waving them on with wild, desperate urgency.

"Jump," Lysera whispered, voice flat with the absolute knowledge of those who have already seen how it ends.

Edrin jumped. The rift yawned wider, jaws ready to swallow, but the jump held. They landed hard, Lysera rolled out of his grip and lay gasping, and Edrin found himself staring up at the weird, blinking storm sky, unable to move for a moment.

Bran was there almost instantly, hands shaking as he helped Edrin to his feet. The lizardfolk's scales smoked from a dozen tiny burns, but his face was alight with the sick joy of a gambler who'd survived the first round of losses.

"You're alive," Bran said. "We're all alive."

Lysera sat up, wiped blood from her nose, and nodded. "Next one is worse," she said, her voice softer than before.

Edrin looked down at himself. His shirt was burned open at the shoulder, blood seeping from the wound, but not too badly. His hammer had survived, though its head was now etched with a spiral of black soot. He checked on Lysera: she looked more shaken than hurt, her skin weirdly bright, her lips blue. Bran's scales had gone dull, patches scorched white by the storm's energy.

Edrin reached for the hammer and hauled it up, the weight grounding him. "Are we there?" he asked, hoping he wouldn't have to do that again.

Bran wiped sweat from his brow, grinning like a skull. "Close enough," he said. "The Path is stable past the next marker. If we move now, we may outrun the storm."

They limped onward. The stitch-storm raged at their backs, but the next stretch of leyline was smoother, the way marked by dull stones and the pale afterimage of whatever had tried to kill them. Edrin found himself almost grateful for the simplicity of the pain: it meant he was still real, still there. Lysera's Sight faded, the glow in her eyes receding, but she held herself together and kept walking, even as the world spun underfoot.

At last, they reached a place where the path blended back into ordinary reality—a patch of half-dead grass and stunted trees, which looked almost boring after what they'd just survived. The sky above was no longer a wound, just a sheet of honest blue marbled with clouds.

Edrin fell to his knees, not out of reverence but because his legs simply gave up. Bran sat down cross-legged on a stone, tail draped over one knee, and began scribbling furiously into his almanac.

Lysera slumped against a tree and stared at the sky, eyes closed. She looked peaceful, in the way only the exhausted ever do.

For a while, none of them spoke. The silence was absolute, a buffer between what they'd endured and whatever came next.

The storm retreated, its flickers now distant and impotent, a memory rather than a threat. In its wake, the three survivors sat, alive and aware that the world had not gotten any kinder, but perhaps a bit more knowable.

#

The first thing Bran did, once he was sure he could trust his own hands not to shake, was uncap his chronometer and start recalibrating. He sat atop a narrow slab of stone—leftovers from whatever trauma had shaped this edge of the leylines—cross-legged and perfectly still, save for the frantic movements of his claws as they flicked between instrument, notes, and a sky-chart that looked older than most gods. Thin wisps of smoke still curled from his scales, their usual luster dulled by the storm's hunger. Bran worked with the grim joy of a beast who'd survived a near-predator.

Edrin paced just below him, boots crunching over shards of wind-blasted stone. His right shoulder throbbed from where the arc had grazed him, but he refused to touch it or show any pain. The hammer at his belt felt heavier now, as if the metal had absorbed some of the storm's violence. Every time he tried to look away from the wounds in the world behind them, his gaze slid right back, seeking assurance that the rift had really closed.

Lysera huddled on a sun-warmed boulder, methodically wrapping her forearms in lengths of clean bandage torn from the lining of her travel cloak. Her hands shook, just enough that she had to tuck the loose end of the cloth under her chin to finish the knot. Still, her eyes were clear, tracking Edrin's restless movement, sometimes flicking to Bran with curiosity or concern.

After a while, Bran clicked the chronometer closed and set it atop his notes. "It's almost beautiful, if you're not being killed by it," he murmured.

Edrin stopped pacing, glowering at the scholar. "What is? The surge?"

Bran held up his almanac, pages crowded with columns of ink and tiny, prismatic pin-marks. "The convergence schedule. I had suspected it was tied to one of the moons' orbits, but the variance was too high for simple astronomy. This—" he tapped the notebook, "—proves the convergence windows are being forced. Someone is nudging the Threadrun, leveraging every instability."

Lysera finished tying off her bandages, then tilted her head. "Nudging? By whom?"

Bran traced the path of a finger down the almanac. "Someone with power and patience. Every time a surge like the one we just crossed hits the leyline, it weakens the boundary. If you time it right, you can create a window for…"

He hesitated, glancing at Edrin.

"… For mantling. For rewriting reality, at least in small doses."

Edrin spat. "So every time one of those storms rips through the world, someone's making it happen? Or making it worse?"

Bran's tail thumped the rock, a nervous gesture. "The storm itself is not artificial, but the intervals—yes, they are guided. Look—" He handed the almanac to Edrin, pointing at a cluster of marks in the upcoming month. "See here? The next Threadrun alignment will overlap with the Solaryn's Crown festival. That occurs in less than two weaves. Every scholar, priest, and mantled bureaucrat in Threadmere will be gathered in one city. If someone wanted to alter the pattern, that would be the place."

Lysera's face paled further, the blue veins at her temples standing out stark against her skin. "And after that?"

Bran shrugged, helpless for the first time. "Then the alignment resets, and the next window is generations away—unless someone keeps pushing the resonance, in which case the world might not survive the stress."

Edrin flipped the almanac back to Bran, his jaw set hard. The injuries to his shoulder and his arms pulsed, faint but real. "Who benefits from this?" he demanded. "What do they gain by burning out the leylines, or rewriting whatever they want?"

Bran's answer was nearly a whisper. "If you could choose your own fate, would you not try?"

Lysera looked up at Edrin, her gaze sharp. "But if everyone else suffers for it, is it still worth it?"

Edrin considered, then shook his head. "Not in my book."

They sat in uneasy silence, broken only by the breeze through the leyline gap and the fading hum of the stitch-storm.

Finally, Edrin spoke. "We need to warn the Mantle Council. They may not listen, but we have to try. If someone's using this magic, all of Aurelia is in the crosshairs."

Bran nodded, already assembling his notes and packing his kit. "We can be in Threadmere before the end of next weave," he stopped to calculate dates, "Maybe on or near the afternoon of Umbersday, Skycrest 20. That's if we use the old markers and avoid the mainline."

Lysera stood, wincing at the pain in her arms, but her voice was steady. "If we're going to be believed, we'll need proof. Something more than bruises and a hunch."

Edrin grinned, despite himself. "Luckily, we've got a lizard with more brains than sense, and a seer who can outstare a god. If the world is ending, at least we'll be in the front, where we can see it clearly."

He hefted the hammer, flexing his right hand until the pain dulled. "Let's move. There's a Council that needs waking up, and I want to be there when their eyes open."

The three of them started toward the city, battered but no longer lost.

#

Threadmere's Mantle Council Chambers sat at the summit of a hill that always looked too clean for the rest of the city. The approach up the white marble stairs was meant to impress humility on the petitioners, but Edrin barely noticed the grandeur.

His eyes were locked on the pair of doors at the top, each one lacquered until the grain gleamed like molten silver. The doors opened at a touch, admitting the battered trio into a rotunda suffused with chill and light.

Lysera paused at the threshold, taking in the interior with a mix of awe and fatigue. Bran, already at ease, stalked toward the reception desk without waiting for the others. The attending clerk, a young man with hair as pale as the marble, glanced up and gave Bran a perfunctory nod.

"You're late," he said, no hint of accusation in his voice. "Councilor Helyas was expecting you at the second bell yesterday afternoon."

Bran's tail twitched. "The storms delayed us," he looked sideways at Edrin and Lysera, chagrined, "I was off by a couple of hours. We have urgent evidence for the Council."

The clerk nodded, producing a wax tablet. "Names?"

Bran provided his. Edrin followed, then Lysera. Each name was inscribed with swift, practiced strokes. The clerk's gaze lingered on Lysera for half a heartbeat—recognition, or perhaps just curiosity at the glyphs reflecting light that wasn't actually there. Then, he gestured them through a side corridor toward the council chamber.

The chamber itself was a semi-circular coliseum, more theater than tribunal, every inch calculated for acoustics and intimidation. Twelve Councilors sat arrayed on ascending tiers, each behind a dais crowded with glyph-etched tablets and silvered ink pots. Between the Council and the petitioners was an expanse of tiled floor, as glossy and empty as a frozen lake.

Edrin stopped at the marker set into the center tile. He could feel the weight of a hundred invisible eyes: some hidden behind the gallery screens above, some just the ordinary pressure of official scrutiny.

Councilor Helyas, a hawk-faced man with silver-threaded hair and a voice that seemed to slice the air, began the proceedings with ritual, "This session of the Mantle Council is called to order. The petitioners will present."

Edrin licked his lips. His tongue tasted copper—old nerves, never far from the surface. He glanced at Bran, who gave him an almost imperceptible nod. Lysera stood at his other side, her hands folded, her wounds hidden beneath fresh linen and a demeanor that projected nothing but calm.

He took a breath, found the center of his chest, and spoke.

"Three days ago, a Loom-surge destroyed Emberfield. My home. My family. My friends. I am the only survivor." Edrin's voice rang off the marble, steadier than he'd expected. "This was not a random surge. My companion Bran—" he indicated the lizardman, "—traced the energy signature. It matches others recorded in Silvertail archives, and the intervals are tightening. Something, or someone, is using leyline resonance to force Threadrun convergences."

Murmurs rippled up the Council. Councilor Helyas frowned, motioned for silence. "You're implying sabotage. On what evidence?"

Bran drew himself up to his full height, scales glittering. "The record is clear. Each event aligns with a weakening of leyline structure in the region. The pattern is not natural. These surges now coincide with major civic festivals, when the population and magical use are at their peak."

Lysera stepped forward. "If the Threadrun is manipulated, the effect could cascade. We might see surges in every city along the leyline. People will die."

A younger Councilor, perhaps no older than Bran, scoffed. "Every year, someone predicts an apocalypse. The Loom survives. We adapt."

Edrin clenched his fists. "The Loom doesn't care if we adapt. It just keeps burning until we're all ash."

At this, a new voice entered—smooth, level, and unhurried. "Isn't it true, Mr. Ashwood, that your family worked for generations in the leyline's shadow? That you might be… predisposed to interpret trauma as conspiracy?"

The speaker was a woman—maybe forty, with composure so absolute it seemed an extension of her skin. She wore a councilor's robe, but her posture and stillness set her apart. She met Edrin's glare with a gaze both appraising and unthreatened.

He recognized something in her face: not a feature, but a quality of control. Lysera recognized that she was facing someone who could not only See, but could maybe even control. Not a woman to be taken lightly.

Edrin drew himself up. "My family lived through dozens of normal glyph and loom surges. I worked on glyphstones, and I know how they can be, at their best and their worst. This wasn't like the others. There was a signature—like a mark. Bran recorded it."

Bran produced his almanac, flipping to the correct page with the flourish of a practiced witness. "Here. The residual is unique, and it matches a spike in Threadmere's own leyline." He passed the almanac up to a waiting clerk, who relayed it to the councilors. The woman in the center—her nameplate read Councilor Vaelis—took the book and began to study it.

She flipped through the pages with slow, deliberate precision. "These are exquisite notes, but correlation is not causation. How do you know the pattern was imposed from outside, and not a natural function of the Loom?"

Bran's jaw tightened. "The frequency. And the precision. It is too elegant for random decay."

Councilor Helyas raised a brow. "So you posit an enemy—a manipulator. Who, in your estimation, has that power?"

Lysera interjected, "There are actors in every city who would benefit. Anyone seeking to rewrite their fate, or amass power, could nudge the Threadrun if they had the right tools."

Councilor Vaelis tapped the edge of the almanac. "And what would you have us do? Suspend the festival, after it started today? Mobilize the Watch? Abandon the city? You realize such an action, if unwarranted, would cause more panic, or even more damage, than the surges themselves."

Edrin fought to keep his temper. "If you wait for proof, you'll be counting bodies. I don't want what happened to my family to happen again. Not in Threadmere, or anywhere else."

Vaelis watched him for a moment longer, then looked to the rest of the Council. "Let the record show the warning has been received. The Council will deliberate."

Edrin almost spoke again, but Bran's hand on his shoulder was a warning. Lysera murmured, "Wait."

Sunlight slanted through diamond-glass transoms, catching on every polished surface. The bench of Councilors arched above the petitioners like a row of celestial judges. Each one was clad in robes stitched with bands of blue or gold, depending on seniority, their faces the color of river clay or obsidian, or in the case of the one with the silver eyebrows, the pale pink of ancient marble. They all watched the trio on the floor with a rather innate detachment.

But it was the woman—centered, poised, utterly at rest behind her station—who finally spoke. Her voice was as level as a plumb line, measured in increments designed for maximum transmission and minimum heat.

"While Emberfield's recent tragedy is… regrettable," she said, pronouncing each syllable as if extracting a splinter from her tongue, "the Council must consider the entire pattern. To divert resources, initiate an audit, or even curtail festival protocol based on a single anomaly would be… rash." She paused, her eyes flicking to Edrin and Bran and finally to Lysera, as if to sweep the field for stray dissent. "We are all aware of the surge schedule and have lived through its consequences. To propose that we should panic now—when the Loom has survived centuries of greater stressors—is to suggest that this generation is uniquely unsuited to the burden of adaptation."

Murmurs swept the council dais, a rustle of parchment and soft agreement. Councilor Vaelis's words hung above the rotunda like incense.

Edrin took a half-step forward, as if to launch himself into the debate again. But Bran's hand, scaly, firm, and suddenly strong, closed over his wrist, stopping him. Lysera stood stock still, her lips pressed into a seam so thin it might never reopen.

The Chairman Helyas raised a hand, palm outward, as if deflecting a wave of invisible force. "The point is well-taken, Councilor Vaelis. The Council will not overreact. However, in the interest of decorum, the petitioners will be granted access to the Archive for a period of three wefts, commencing. Any data, any proof, that emerges during this time will be reviewed with urgency. Until then, no further action will be taken."

Councilor Vaelis inclined her head a precise degree. "Prudent," she intoned. "If there is substance to these claims, let the Archive be the judge."

A clerk in gold-trimmed black appeared at Edrin's side, gesturing him toward the exit with a hand that would have seemed polite anywhere but here. "You may retire to the anteroom. Arrangements for Archive access will be made within the hour."

For a second, Edrin looked ready to argue—his whole body a series of taut lines and unfinished threats. But Bran steered him firmly toward the doors. Lysera moved after them, her step light as air, but her shoulders set in a way that suggested collapse was not far off.

The doors closed behind them with a hush so complete it felt like entering a tomb.

#

Outside the council chamber, the corridors were a different world: sunlight diluted to gentle amber, the hush broken only by the distant click of a timekeeper's shoes or the fluting cadence of a passing page. The three of them drifted down the first flight of stairs, then another, and finally slumped onto a stone bench half-hidden by a potted evergreen. Here, the world resumed its regularity: the distant bell of a newsboy, the sound of children in a far-off courtyard, the low drone of a city busy not with emergencies, but with the steady, endless business of staying alive.

Edrin sat forward, elbows on knees, staring at the veins in the marble tiles. When he spoke, his voice had lost its edge, gone blunt from collision. "Did you see how they looked at us? Like we were ghosts there to haunt them. Like we'd tracked in something dirty on their nice floor."

Bran's reply was a dry click. "They have lived through centuries of surges. To them, we are not the warning. We are the disruption. But," he added, and his hand closed tighter around his battered almanac, "the Archive holds the real evidence. If we can find the match, or the vector, they will have no choice but to act."

Lysera folded her hands in her lap, fingers laced so tightly the tips turned white. "They won't act, not unless the Loom forces them. Or unless we do something so unignorable that they lose the privilege of denial." She looked at Edrin. "Are you willing to go that far?"

Edrin's answer was a barked laugh, short and without joy. "You heard what's left of my home. There's nothing left to lose, not for me." His gaze flicked to Lysera's bandaged arms, thinking of how dangerous the road ahead could be. "But you don't have to come with us. You could stay here, let the world fix itself without our help."

She shook her head. "The Loom doesn't fix. It only persists." Then, softly, "If I stop now, the visions will kill me by inches. I'd rather try for something better."

Bran watched them both, unblinking. "Then it is decided. We go to the Archive."

They rose, moving as a single organism now: three cells in a chain, each linked by necessity and pain and a small, shared hope. They walked past the mural of the first Mantle Conclave, past the statue of a previous councilor who had once outlawed open flames in the city, past the echo of the council's own indifference. At every turn, they were shadowed by clerks and custodians, none of whom met their eyes for long.

They reached the outer doors, a pair of iron-and-glass slabs designed to suggest permanence and clarity. Lysera opened one, letting the world's brightness pour in around them.

Just before they stepped through, Edrin turned back, as if some muscle in his spine refused to let him go. At the far end of the corridor, she stood: Councilor Vaelis watching them with an expression as unreadable as the moon. There was no malice there, nor triumph, only an infinite patience and a purely intellectual hunger.

Their gazes met for a single beat of the heart.

Vaelis's hand rose, as if to wave, but the gesture was so minuscule it could have been a flinch. Or a warning. Or the start of a new thread, winding out from the old.

Edrin held her gaze. For the first time since Emberfield burned, he wondered whether he was winning or losing, and if he even knew what game was being played.

Then the door closed behind him, and the world became noise and light and the promise that maybe, just maybe, a single voice could matter.

#

Inside the Council chamber, the councilors resumed their positions, their conversation already shifting to budgets, parades, and the logistics of a thousand-year-old city trying to remake itself in the new era. Vaelis sat perfectly still, her hands folded in front of her, her eyes never leaving the door that Edrin, Lysera, and Bran had just passed through.

She could still feel the afterimage of the hammer-wielder's rage, the shimmer of the seer's magic, the crackling possibility of the lizard's insight. She let it settle, then folded it up inside her mind, a gift-wrapped hypothesis.

No one would remember her name. That was as intended.

But the world would remember what happened next.

CHAPTER FIVE

Road to Verdantia

Chapter Five

Road to Verdantia - Afternoon (Morday, 21 Skycrest, TT 1215)

The first warning of the attack was a flicker of metallic light, gone as soon as it caught the eye. A glint off steel, or maybe the sun catching a bit of polished mirror, high on the ridge above the old Verdantia road. By the time the second warning—three sharp bird calls, too regular to be real—reached the ears of the merchant caravan, the sand-hounds had already spooked, and the first wagon's wheel was a casualty. There was a snap, a warbling scream, and then the world descended into a symphony of falling wood, trampled goods, and the shrill cries of opportunists who'd waited weeks for a haul like this.

Bandits emerged from the scrub with a kind of desperate choreography. They were not the uniformed, organized raiders of the old Mantle Wars, but something shabbier, all patched trousers and hastily-scavenged armor. They shouted conflicting orders and brandished mismatched blades, each man or woman hoping to be the one with their hands deepest in the loot when the dust settled.

The lead wagon, blue paint blistered by the sun, tipped as two masked figures pried its axle with a crowbar. The driver—a merchant of the fussy, well-fed sort—piled out the back, clutching a ledger and a sack of coin. Someone winged a crossbow bolt in the general direction of the attackers, but it missed by a distance measured in city blocks. All around, travelers and guards abandoned dignity in favor of survival, diving into the powdery roadside or, in one memorable case, behind a basket of cabbages.

It would have been a predictable massacre, if not for a sun-browned man in a tattered red bandana, grinning like a fey fox with feathers in his teeth. He came charging down from the ridge with the bored, easy confidence of someone who had already calculated every possible outcome and found most of them amusing. The bandits spotted him at the last second—just long enough for one to croak, "Not again!"—before he was among them, a sword in each hand and a motion that was more dance than duel.

"Ah!" the stranger's boisterous, low-toned voice cried out, "You must have faced me before!" He cut down the first with an almost gentle, horizontal slash, as if persuading the man to lie down for a nap. The second, he dispatched with a spinning kick, sending the unlucky soul sprawling into the shattered wheel. The third, a woman with a rusty trident, managed a single lunge before the red-bandana man sidestepped and rapped her on the helmet hard enough to just ring her bell. "It's odd. Most who have faced me before are dead."

His movements were showy, full of flourish and wasteful energy, but there was never a moment when it looked like he might lose. Between every stroke and parry, he flashed the whites of his teeth or threw out a theatrical wink, as though he expected someone, somewhere, to be keeping score.

He wore a battered leather vest and a set of blousy silk pants that contradicted the deadly nature of his attacks. The vest had something embroidered on the back in fading gold thread, but the man was moving too fast to read. The look was somehow less comic than it seemed. There was something about the man's presence that made the clothing seem less like a joke and more like an old scar, hard-won and borne with pride.

On the hillside above, Edrin Ashwood watched this entire performance with a mix of incredulity and ill-disguised envy. The bandit attack had only just registered before the newcomer had swept in, leaving Edrin and his companions, Bran and Lysera, little to do but observe. In Edrin's case, he clutched the handle of his father's hammer until his knuckles turned bone-white. They happened upon the attack as they had crested the hill while they were back on the road to explore a new lead that Bran had found in the archives of Threadmere about the surges and convergences.

When the dust finally settled and the last of the would-be thieves either fled or lay unconscious, the warrior strode up to the toppled wagon and offered the merchant a hand up. The merchant took it, eyes wild, and babbled thanks that were at least fifty percent promise of future discounts and another fifty percent religious gratitude to whatever god had sent him.

"Always happy to do a little crowd control," the stranger said, voice as bright and sharp as the blade he twirled absentmindedly between his fingers. He righted a crate of salt fish with a boot, then turned his attention to Edrin's party on the hill as if only now noticing them.

"Company!" he called, flashing a salute. "Looks like you missed the fun, but there's plenty of clean-up left."

Edrin, after a beat, started down the slope, Bran and Lysera close behind. He felt awkward—three days in council chambers had done nothing to dull the urge to hit something, and now the opportunity was gone. As he reached the road, he hefted the hammer in a show of solidarity, or perhaps challenge.

"Didn't want to take your glory," Edrin said, grinning despite himself. "In fact, I don't really know much about fighting. I just hit things with this."

Malrik's eyes went to the hammer. "Family heirloom, I'll wager?" He stepped forward, extending a hand not for a shake, but for the old forgeman's grip, wrist to wrist. Edrin accepted.

"Edrin Ashwood," he said. "Last of my line, apparently."

"Malrik. No line to speak of, unless you count horses and heartache." He winked again, the expression infectious. "But you can call me the Gypsy Knight, if you want to embarrass me in public." As he turned away, Edrin could almost read the High Aurelic script on the back of the man's vest. He had never been very good at lessons before he apprenticed as a smith. The closest that Edrin could determine was "Knight of the Wanderings."

Bran sized up the situation in two heartbeats and began circling the perimeter for stragglers. Lysera stood back, eyes on the still-twitching bandits, hands clasped in front of her.

Malrik surveyed the ruins of the caravan, kicking an unconscious body nearby. "They get bolder every year," he said. "Or maybe just hungrier. I'd hoped the roads were safe out this way, but hope and reality rarely shake hands."

Edrin eyed the merchant, who was now crawling under his wagon, presumably in search of his ledger. "Looks like you saved their hides."

Malrik shrugged, but it was a theatrical gesture. "If only I could save them from their own bookkeeping." He stooped to pick up a fallen trident, inspecting it with a craftsman's eye. "So, what brings you three to these forsaken flats? Not the scenery, I hope." He gestured to the scrub, flat fields and rolling dunelike hills.

"We're headed west," Bran said, rejoining the group. He kept a cautious distance, as though Malrik's confidence was more dangerous than any weapon. "Research. Archive business."

"Ah, Threadmere folks!" Malrik said, slapping the trident against his palm before tossing it onto a pile of discarded weaponry. "I figured you two—" he pointed to Bran and Lysera, "—for academics. The careful way you walk, the way you scan a crowd. Not like the usual merchant muscle."

Lysera smiled, but it didn't reach her eyes. "We're just passing through."

Malrik's gaze lingered a little too long on her, then flicked to Bran, then back to Edrin. "Well, then. Let's not stand on formality." He stepped aside, gesturing with both hands at the battered wagon. "Help yourselves to a drink, if there's anything left unbroken. I'll handle the dead weight."

He began dragging the unconscious bandits to the side of the road, arranging them in a neat row. Edrin, after a moment's hesitation, moved to help him. They worked in silence for a bit, Edrin noting the way Malrik checked each bandit for signs of life before depositing them on the pile. Malrik didn't stop at searching them for signs of life; he helped himself to any glyph coins or other trinkets the dead had on their bodies.

Once finished, Malrik produced a flask from somewhere in his vest, took a swig, and offered it to Edrin. "To surviving another day on Aurelia's worst highway," he said.

Edrin accepted, surprised by the clean taste—apricot and smoke, a hint of something herbal. He coughed, then grinned. "Not bad."

"Better than dying of thirst," Malrik said, then leaned in, voice lowering. "You watch the council's business, you see the shape of things to come. There's more of this chaos, folks scrambling for scraps, every month. I think the Loom's getting hungrier. Or maybe just meaner."

Edrin studied him, not sure what to make of the statement. "You think the state of things is deliberate? Not just bad luck?"

Malrik shook his head. "Luck is for gamblers. This—" He gestured to the fallen bandits, the ruined wagon, the horizon beyond. "—is the slow unwinding of something old. World's changing, and not for the better."

He smiled then, broader than before, as if to break the tension. "But for tonight, we eat and drink. The future can wait its turn."

Lysera had found a seat on an overturned crate, her hands busy mending a small cut on her palm. She watched Edrin and Malrik, measuring their every word and motion. Bran, for his part, had moved to help the merchant with the axle and was already outlining a plan for repairs.

The sun slipped toward the horizon, turning the dust to gold. The worst of the day's violence faded, replaced by the communal ritual of fixing what could be fixed and forgetting, for a little while, what could not.

As the wagons were righted and the last of the goods restacked, the two parties gathered by a makeshift campfire—sad and still a little anxious, most managed a cheerful smile as flames licked at dried brush and wagon splinters. While some of the merchants produced some root vegetables and some dried meat, and began a stew, Malrik produced a battered concertina from the depths of his pack and played a tune that was part lullaby, part drinking song, part lament for the world as it used to be.

Edrin found himself humming along, Lysera quietly harmonizing, and even Bran joining in for the chorus.

For a while, there was peace—enough of it to pretend, at least, that the world could be mended if only people tried hard enough.

After his song, Malrik stood, stretched, and dusted off his vest. "Tomorrow we part ways, I suppose," he said, his voice uncharacteristically soft. "But I have a feeling the road's not done with us yet." And he moved off to relieve himself by a nearby bush.

Edrin nodded, feeling the truth of it somewhere deep in his chest.

#

After the last of the makeshift stew was scraped from the cookpot and the day's wounds bandaged with rough cloth and good-natured curses, the two groups merged into a single, fire-lit organism. They ringed the flame in deliberate disorder: the merchants hunched in a tight knot around their ledgers, while the travelers—Edrin, Bran, and Lysera—sprawled where the warmth reached deepest. Malrik claimed the spot nearest the blaze, his boots caked with dust, bandana damp with sweat, the gold threading on his vest now obvious in the flickering firelight.

He wasted no time filling the air with his own voice. Malrik spoke with the cadence of a man used to competing with wind and thunder, every sentence a punchline or a preamble. His tales were half-lies and whole adventures, punctuated by wild gesticulations and the thump of his wooden cup on the upturned crate that served as a table. "There was a time I fenced with a mantled baroness in the back alleys of Threadmere," he said, "and lost only because she'd sewn steel into her corset, the cheating witch. Cost me three teeth and a perfectly good pair of trousers." He grinned, displaying a full set of teeth—none missing, all in perfect, predatory alignment.

The merchants soaked up every word, eyes wide and lips moving in silent calculation: how many coins or curses would it take to turn this vagabond hero into their permanent bodyguard? Edrin, on the other hand, watched the man with a blacksmith's scrutiny, as if waiting for the crack in the blade, the flaw in the steel that would reveal whether Malrik was legend or just another liar. He also didn't like the way that the man seemed to covet Lysera every time he looked at her.

Even when he laughed—especially when he laughed—Malrik's eyes never truly left the perimeter. He drank in the fire and the faces, but always with a part of himself reserved for the darkness beyond, measuring every shadow for movement. Edrin noticed and approved. It was the same vigilance that kept people alive on the forge floor or in the ruins of Emberfield.

Bran, for his part, seemed immune to Malrik's charm, but only because he was too busy interrogating the specifics of every story.

"You said the baroness wielded a blade of unbreakable glass. Was it an artifact of the Mantle Wars, or a product of the later Schools?" he'd ask, eyes glinting with the possibility of a new data point. "And the catacombs in Sunpeak—did they truly spiral down thirteen levels, or was that a symbolic exaggeration?" Malrik never tired of answering; if anything, he embellished further, always careful to wink at the gaps between truth and performance.

Lysera stayed mostly silent, hands folded in her lap or tracing subtle circles in the dirt beside her. Once, when Malrik's stories turned to the darker territories—lost cities where the Loom had devoured every scrap of color and sense—she flinched, just barely, unnoticed by everyone but Edrin. He wondered what she Saw, and whether she would ever say it aloud.

As the fire dwindled to a low, sullen ember, and the merchants faded off to their wagons to sleep beneath stacks of salted cod and account books, the conversation shifted. Bran, perhaps emboldened by fatigue, steered the talk toward the rumors that had sent them west. "You've seen Loom-surges, then," Bran said. "Not the minor ones—the true anomalies. The ones that leave a town changed, or burning."

Malrik's expression shifted. The smile was still there, but now it was less a performance and more a confession. He leaned forward, elbows on knees, voice dropping so that only those still at the fire could hear. "You see enough surges, you start to smell them before they start. The air goes tight, like a drumskin. Static on your teeth. Then it's over, and the world doesn't remember what it was supposed to be. Only the dead remember, and not for long."

Edrin nodded, remembering the taste of ash and ozone in Emberfield, the way the light went flat and sour before the surge hit.

Malrik's gaze shifted to Lysera, as if sensing her reticence. "You're quiet, Seer. Haven't you got a riddle or a warning to drop in the pot?"

Lysera looked up, her eyes black in the shadows. "Not all visions are fit for sharing," she said. "Sometimes the pattern is clearest when you leave it unspoken."

Malrik considered that, then laughed—not unkindly. "Good answer." He drained his cup, set it down with finality, and ran a thumb along the rim as if testing for cracks. His hands, Edrin noticed, were scarred and callused in patterns speaking of more than just swords—rope burn, acid blisters, the kind of damage that came from working odd jobs in bad places.

The night had grown cold, the wind picking at the last bits of warmth. Only the three travelers and Malrik remained at the fire now; the circle shrank to its honest shape.

Bran pressed on. "If you sense Loom anomalies, maybe you can help us. We're tracking a pattern, something deliberate, not just random collapse. Have you seen surges that repeated, or left marks the way a ritual would?"

Malrik's head tilted, the smile slipping for a heartbeat. "What kind of marks?"

Bran reached into his kit, produced a slip of parchment, and traced a spiral glyph in the dirt with a bit of burnt stick. "This," he said. "The signature recurs at every site. Sometimes in stone, sometimes just burned into memory."

For a moment, Malrik stared at the spiral, unmoving. His thumb, still on the cup, began to tremble—so slightly that only Edrin would have caught it. Then his whole body snapped back to motion, and he clapped Bran on the back, nearly spilling the scholar into the embers.

"You're persistent, I'll give you that," Malrik said, but the words now carried a hollowness, as if rehearsed. "I've seen that mark, yes. In ruins, on burned villages, in the eyes of men who wake screaming and never remember why. The old hands in my trade call it the Threadbreaker's curse. Most won't even speak of it."

Lysera's hands went still. Edrin's grip tightened on his hammer. The silence that followed felt loaded, as if even the wind paused to listen.

Malrik rose, dusted off his trousers, and stretched. All the old bravado returned. "But you lot have the look of people who won't be warned off by a curse, so I'll give you free advice." He pointed west, toward the dying embers of the sun. "Keep walking. The spiral always leads to the same place. And I think, if you're unlucky, you'll find out where."

He bent to gather his kit—a roll of battered maps, the concertina, a pouch of dried fruit. He spoke without looking up. "I'd say it's been a pleasure, but I know better than to lie at a campfire. Try not to die before you see the ocean."

Edrin opened his mouth to protest, or maybe to ask Malrik to join them, but before he could form the words, Malrik had already stepped into the darkness. His boots crunched the hardpan only twice before the sound vanished.

Bran was the first to break the spell. "He left too fast. Like something was chasing him."

Edrin stared at the spiral in the dirt, now blurred by Malrik's boot. "Or something calling him," he muttered.

Lysera's fingers resumed their tracing, drawing new circles in the ashes.

A few minutes later, the sounds of footfalls approached again, and Malrik appeared out of the darkness, smiling his big smile. "You thought I was gone. I just wanted to know what you would say about me behind my back." His grin continued to show, "I am not going to leave a fire and some trustworthy companions and walk the road at night. And I *always* lie at a campfire." Malrik unrolled a thin blanket and then, of all things, a small pillow and settled in next to the fire to get some sleep.

Bran looked over at Edrin and mouthed, "What was that all about?" Edrin simply shrugged and looked back into the fire.

The night closed in, the warmth of the fire fading, and for the first time in weeks, Edrin felt the return of the old, familiar dread: that the world was much bigger, and much more dangerous, than anyone could understand.

He lay down on the bare earth, staring at the stars, and wondered if any of them would ever write a song about him. It wasn't as if he knew what he was doing. None of them did. They just knew that they had to keep moving to try to find out what could happen and if Edrin ever found that there was a person, god, or creature behind what had happened to him, and his world, the Loom itself couldn't protect that thing from Edrin's wrath.

#

Moonlight spilled across the marble floor, catching on the jagged edge of the mirror, stretching from floor to ceiling. The mirror mage stood before it, bare skin shimmering with a faint, unnatural luster, eyes fractured into a dozen reflections. Her fingers traced the mirror's surface, painting invisible glyphs that shimmered silver and then vanished. Her naked body was weaving back and forth across all of the mirrors. She could see her ample breasts, with reddened tips pointing upwards in one mirror, and by shifting her head to look at a different mirror, she could see her legs and equally ample backside reflected. Her mouth broke into a sensual smile, like she was teasing herself. "Enough," she whispered to herself, "time to work."

She focused, calling forth the image she desired. Glass rippled, swirling into visions: First, the feeble caravan fires—Edrin, hammer across his lap, Lysera and Bran beside him, all wary and wounded. Next, Malrik's face, bold and careless, grinning as if he'd already won every contest the world could offer.

The witch's lips parted in a cold, hungry smile. "So much will. So little wisdom." Her voice echoed, layered over itself, as she stepped closer to the glass, letting the images sharpen. She watched Edrin for a long moment, tracing the pulse of anger and hope in his aura. But it was Malrik who drew her, the way his presence burned at the edge of order, reckless and volatile.

"You'll do," she whispered, breath fogging the mirror. "You'll be my vessel. My mask. My sword." She pressed her palm to the glass—naked, vulnerable, yet utterly without shame. Mirror-shards shimmered around her, each showing a different path:

Malrik conquering, Malrik burning, Malrik kneeling in blood. Malrik forcing himself on a young woman.

She drew a glyph across the reflection of her own chest, the lines reflecting and refracting until they stitched themselves into the silver surface of the closest mirror. "Now, we begin."

#

Malrik slept sprawled beneath the stars, one hand on his blade, the other fisted in the dust. Dreams crawled in through the cracks of his mind, cold and sharp as broken glass.

He stood at the mouth of a ruined kiln, flames flickering blue-white and roaring without sound. A shadow stepped from the fire—a woman with eyes of shattered mirror, her skin made of moonlight and wounds. She beckoned, voice ringing through his bones: "You hunger for power, don't you? For meaning? The fire god's relic waits. Take it, and all the world will burn for you."

Scenes spilled out in quicksilver flashes: Malrik with a crown of embers, his skin peeling in molten sheets. People bowed at his feet, or cowered, faceless in the blaze. He reached for the relic—an orb of blackened glass and iron—and felt a rush of heat, pain, and a tether pulling him toward purpose.

He awoke with a gasp, sweat freezing on his skin despite the stored desert heat still radiating from the ground. He realized that his manhood was straining against his underclothes, rigid almost to the point of discomfort. The camp was quiet, the others asleep or pretending. Malrik sat up, heart hammering, the echo of the vision burning in his mind.

He looked west, toward where the sun would rise, and grinned. For once, he whispered, "A fire relic, is it? Well. Can't let opportunity pass." By dawn, he was already a league away, footsteps silent as smoke. He left behind only the memory of a song and the promise of fire.

#

After Malrik's early morning departure, Edrin, Lysera, and Bran tried to help the merchants get everything back on the road and moving slowly down it. Walking with a slow-moving caravan gave too much time for thought and remembrance for Edrin's taste. But he was a quiet man anyway, so he didn't suffer with little to no conversation heading towards him. He could keep his eyes constantly on the hills and roads for signs of another raid.

On their second night, after the stories and the stew, the cast-together group tried to sleep around some fires. As the fires died and people stopped talking quietly, wind and moonlight visually scraped the campsite to black and white, every object looking like bones. Where there had been music and laughter, there now remained only a trio of orange-glowing knots and the soft, irregular breaths of people trying, and failing, to sleep. The merchants had burrowed into their wagons. Edrin and Bran kept to the lee side of a half-burnt log, their backs to the dying warmth, while Lysera curled into herself, face hidden by a cascade of midnight hair.

So, it was not the fire, nor fatigue, that woke Edrin when the shadow shuffled into their circle—it was the smell. Resin, old blood, and the acrid sweetness of burnt prayers. The figure was a skinny boy of twelve, carrying a coil of rope more twisted than any Edrin had seen at the forge.

He crept closer, hands trembling, and stopped just outside the ring of fading embers. "Are you the one with the hammer?" the boy whispered. His voice was pitched high, with the careful diction of someone who wanted to be sure he was not misunderstood.

Edrin sat up, blinking hard. "I am. What's the trouble?"

The boy looked over his shoulder as if expecting to see a parent, then darted into the firelight and thrust the rope at Edrin. "This. The Gypsy Knight bought six of them, but said it wasn't enough. He told me to tell you, if I could find you, that he's 'leaving a trail for the stubborn ones.'" The words were half-rehearsed, the cadence of a message carried by a child for the promise of a coin, or maybe just a new story.

Bran rolled over, eyes already awake. "What do you mean, 'not enough'?"

The boy shifted his weight, uncomfortable. "He said there were more kilns in the next two towns, and he needed them all. He asked which ones 'breathed, not burned,' and he paid with a coin I can't spend. Look—" He fished in his pocket, produced a disk of some off-white metal, heavy and cold-looking even in the moonlight. "Said it's real, but it's not for money."

Bran took the coin, rolling it between his fingers. It was stamped with a spiral so fine it seemed to move under his touch. "Pyrrhos," he said, showing it to Edrin. "The old fire cult. They haven't minted these since before the Mantle Wars."

Lysera roused herself, eyes shining in the low light. She took the rope from Edrin, turning it over, her hands careful. "This is votive," she said. "Used for warding or sacrifice. The pattern—"

she tapped at the way it looped and rejoined itself, always returning to the center, "—is for containment. But this isn't the right thread. It's not even from the right plant. It's… not right."

Bran's scales dulled as he reached for his own kit. "Let me see." He took a pinch of the gray from the knot, scraped it onto a glass slide, and held it to the last scrap of firelight. The ash glittered—not with the soft, chaotic sparkle of wood or bone, but with a hard, vitreous glint. He produced his small prism and passed the flecks through its light. For a heartbeat, the rope's dust burned with a blue flame, then went black, leaving behind a smear that caught the dying coals and reflected them as a perfect miniature sun.

Bran's mouth worked silently, then: "Mirrathar," he gasped, whispering. "That's mirror-dust. Residue from a mirror mage's workings."

Edrin's stomach clenched, a gut-deep revulsion he could not explain. "He's not just lighting fires, is he?"

Bran shook his head. "No. He's making a fuse."

The boy was edging away now, duty done, but Edrin caught him by the arm. "Did Malrik say where he was headed next?"

The apprentice hesitated, then nodded. "He said to follow the smoke. And… and he said you'd know the number before the end." He broke free and vanished into the night, leaving behind a smear of soot and the hollowed-out silence of bad omens.

Bran set the coin down, careful not to touch it with bare skin. "If he's priming a ritual, it will happen on a cycle. Seven days to bank the coals, two more to starve the air, then ignition. It's a combustion cascade, but for people, not fuel."

Lysera's eyes had gone glassy, her voice an echo of someone else's. "He's not alone," she said. "He counts aloud, but there's a woman's voice behind his. Like a teacher in a dream."

Edrin was already gathering his things, knotting his boots, wrapping the hammer handle with rawhide for better grip. "We have to stop him before nine, then. If we wait, it'll be too late."

Bran pocketed the Mirrathar coin with a pair of tweezers, then slung his kit onto his back. "What about the others? If the kilns in the next town are already banked—"

Edrin didn't let him finish. "Then we find the last one. If we break the pattern, maybe we break the ritual. Or at least slow it down."

Lysera stood, brushing dirt from her robe. Her hands were steady now, her expression hard. "Let's go."

They broke camp in silence, leaving the fire to die behind them. The last embers hissed as the wind picked up, carrying the scent of resin and the faraway promise of a new, even more dangerous dawn.

The road west was a line of blackness, the moon a ghost at their backs. Somewhere ahead, the Gypsy Knight was counting. Every step Edrin took was a silent answer: one, two, three. And always, always, the threat that the spiral would close before they got there.

They walked as fast as they dared, and the world, behind them, burned a little brighter with every stride.

Chapter Six:

Road to the Kiln Fields

Chapter Six

Road to the kiln fields - Dawn (Thalday, 24 Skycrest, TT 1215)

Dawn found them exhausted, the night's march etched into their faces like weathered glyphs. Edrin's shoulders slumped beneath the weight of his hammer, Bran's scales had dulled to the color of old pewter, and Lysera's eyes were rimmed with shadows as deep as her visions. None of them noticed the figure on the ridge—still as ancient stone, a silhouette against the paling sky, watching with the patient focus of a predator that had finally found its prey.

He descended with the grace of shadow falling, boots finding purchase on scree that should have announced his approach with betraying slides and tumbles. Instead, he appeared before them between one step and the next, as if he'd always been there, waiting for them to notice.

"The hammer-bearer walks a smoking path," the stranger said, his voice a low melody, twisting around itself. "The scholar seeks what burns, and the seer counts breaths not her own."

Edrin's hand dropped to his hammer, but Bran's reaction was entirely different. The lizardfolk's head snapped up, his dull scales suddenly alight with recognition.

"Theryn," Bran breathed, the name carrying a weight of memory. "It's been years."

The newcomer's eyes were a striking, deep green, flecked with gold, catching the rising sunlight, and reflecting it like captured flame. His face, framed by a wild cascade of dark brown curls, held the kind of beauty that made one uneasy—too sharp, too knowing, too aware of the world's hidden mechanisms. Intricate tattoos snaked up his forearms, each symbol telling a story of deception and truth, the ink seemingly alive against his skin as his muscles moved, as he extended his hands in greeting.

"Time passes differently for those who walk between stories," Theryn replied, a smile playing at the corners of his mouth. "But some tales demand witnessing."

Edrin glanced between them, wariness evident in the tightening of his jaw. "You know this man?"

"More than know," Bran said, his tone softening with rare emotion. "When my clan exiled me for seeking forbidden knowledge, it was Theryn who found me half-starved in the marshlands. He guided me to Threadmere when I was too young to find my own path." The lizardfolk's scales shimmered with remembered gratitude. "He showed me that knowledge could be a refuge when family was not."

Theryn's gaze shifted to Edrin, assessing him with the calm scrutiny of someone accustomed to measuring truth. Theryn patted his pants pockets, then the pouch nearest his shoulder. He extracted what looked like a dried root and bit it between his teeth, before pulling out a large match and lighting the cheroot and sucking the smoke into his lungs. "The last Ashwood carries more than his father's hammer," he observed. "He carries a question that burns hotter than any forge."

Lysera stepped forward then, her exhaustion momentarily forgotten. "We've met before," she said quietly. "In Veilwood

Hollow, during the small eclipse three winters past. You were trading stories for secrets at the Whispering Glade."

Theryn's smile deepened, revealing a glimpse of genuine warmth. "The seer remembers. Few do when I pass through their lives. Fewer still recognize the importance of that meeting." He turned to address them all. "I've been tracking the same footprints you follow—ash-paste and mirror-dust, prayers that burn the tongues that speak them."

"Malrik," Edrin said, the name uttered like a curse.

"The Gypsy Knight is merely a vessel," Theryn replied, his voice dropping to a near-whisper. "The hands that guide him belong to another. I've watched him gather components for the ritual—seven nights to bank the coals, two more to prepare the sacrifice. Tomorrow, he spills blood."

Bran stepped closer. "You've seen the pattern too? The spiral glyphs at the sites?"

"What is seen with eyes is often the least important part of seeing," Theryn answered, his fingers tracing invisible patterns in the air. "The nomadic stones speak of older paths beneath new ones. The Whisper's lineage remembers what others choose to forget."

"He means yes," Bran translated, a hint of fond exasperation in his voice. "Theryn's people are keepers of deception lore. They study illusion not to practice it, but to pierce through it."

Edrin's hand remained on his hammer, but his stance had relaxed slightly. "And why come to us now?"

"Because three paths converge stronger than two, and four stronger still," Theryn said, pointing to each of them in turn. "The hammer that breaks, the scholar who reads, the seer who witnesses, and now the riddleweaver who unveils." His expression grew serious, the playful edge of his words falling away. "What Malrik builds is not meant for this world. It tears at seams already fraying."

Lysera nodded, recognition in her eyes. "The voices I hear when I count—they're real, aren't they? Someone is speaking through him, guiding his hands."

Theryn's gaze locked with hers, smoke curling lightly from his nostrils. "Two threads entangle—his voice and another's. The second speaks with honey and mirror-light, weaving illusion so deep even the weaver half-believes it."

"Mirrathar clan," Bran murmured. "The mirror-dust confirms it. Someone from that cursed family must be trying to revive old magic that we don't remember or haven't found in the archives."

"If you're offering help, we accept," Edrin said bluntly, fatigue and urgency stripping away any diplomatic pretense. "But we move now. The trail's already cold."

Theryn smiled, something ancient lurking behind his eyes. "Cold for those who track with eyes alone. For those who read deception's wake, no trail ever truly fades." He gestured toward a line of low hills half-lost in morning mist. "The Gypsy Knight passed through Ashvale toward the Red Kiln. His steps are heavy with purpose, but light with another's guidance."

Without further discussion, the now-quartet set off, the addition of Theryn subtly altering their formation. He moved with fluid ease beside Bran, occasionally murmuring observations that only the lizardfolk could comprehend. Yet his presence added a current of certainty to their pursuit—as if the addition of this riddleweaver had completed some necessary equation.

As they walked, Edrin fell in step beside Lysera. "You trust him?" he asked, voice low.

She watched Theryn's back, the rising sun caught in his dark curls. "I trust what I saw in Veilwood. He knows truths and lies intimately, but he chooses truth when it matters." She paused, something like wonder crossing her features. "He showed me how to quiet the visions when they become overwhelming. Few outside the Hollow understand that skill."

"Can a Seer's visions be that bad?" Edrin asked, finding that he was curious about this young woman, who, somehow, only a threadturn or two younger than him, exuded a maturity that he did not.

"Oh," Lysera said solemnly, "I thought you knew. Most Seers die young and usually at their own hands. The visions can be so raw, so savage, so…" She stopped her steps, struggling for words.

Edrin cut her off, stopping on the path and lightly holding her hand, "Don't. I asked a flaming stupid question. You don't need to worry. We are together as a group. You are protected here. Nothing will harm you while we are around."

Lysera smiled and pulled her hand out of his. She hoped that he hadn't noticed how tightly she initially held his hands. They began walking to catch up to Bran and Theryn.

Ahead, Theryn turned, as if sensing their conversation. The morning light caught his green-gold eyes, making them flash with an almost supernatural brightness.

"Every deception leaves a trace," he called back to them, "just as every truth leaves an echo. We follow both to find our quarry." His smile was both invitation and promise. "The Gypsy Knight thinks himself clever, but no mask is perfect when scrutinized by those who understand its making."

And with that, he turned forward again, leading them toward whatever future waited beyond the misty hills. A future that, with his arrival, seemed both more complicated and more possible than before.

#

Pyrrday, 26 Skycrest brought a sky bleached white with heat, the air shimmering above the dusty trade routes webbing the region like ancient, neglected veins. The group split at dawn, each taking a different approach to the scattered waystations where information flowed freely. They hunted for whispers of Malrik—a flash of red bandana, the echo of a concertina, the lingering scent of mirror-ash following his passage like a ghost's shadow.

Edrin found the kilnmasters at the eastern waystation, their hands permanently stained by the soot of their trade. Their fingers were cracked like old leather, nails rimmed with black that no amount of scrubbing would remove. He approached them with the directness of a hammer strike.

"A man passed through here," he said, setting his own hammer down with deliberate care. "Calls himself the Gypsy Knight. Charming, deadly with a blade, interested in your kilns."

The eldest kilnmaster, a woman with wrinkles and burn scars, spat into the dust. "That one," she said, the words carrying the weight of suspicion. "Paid in odd coin for a look at our deepest chambers. Wanted to know which ones still held heat when the fires were out." She gestured to a brick chimney that leaked a thin trail of blue-gray smoke. "Left that burning hotter than it should. Something ain't natural in the way it draws air now."

Edrin leaned closer. "When?"

"About three days passed. Left before sunrise, heading northwest toward the old temple grounds."

Meanwhile, Theryn wove through the market square, his approach as indirect as Edrin's was straightforward. He became a shadow amongst shadows, a listener at the edges of conversations, his questions buried in seemingly innocuous exchanges. A copper coin for a cup of ale bought him a seat beside a wagon driver whose eyes darted nervously toward the hills.

"The road to Red Kiln grows warm underfoot," Theryn observed, sipping his drink. "Almost as if the earth itself has taken fever."

The driver glanced around before leaning in. "A strange man came through asking about the old breathing stones. Not many remember those—places where the rock itself seems to inhale and exhale with the cycles of the moon." His voice dropped further. "He left offerings at each one. Not the normal kind. Something that burned blue and left the air tasting like metal."

Theryn nodded, his expression betraying nothing while his mind catalogued every word. "Some say such places are doorways," he replied.

"Doorways should remain closed," the driver muttered, draining his cup and rising to leave. "Especially when something's knocking from the other side."

Across the waystation, Bran approached matters with scholarly precision. He donned his glyphed glasses and examined kiln structures with the detached fascination of an anatomist, measuring dimensions and calculating heat signatures with instruments pulled from his ever-present satchel. The quarry foremen wiped perpetual sweat from their brows as they answered, pointing to disturbed amphorae of oil—seals broken by hands with no right to touch them.

"The man asked about the old vent systems," one foreman explained, his bulk casting a welcome shadow over Bran's notes. "Wanted to know which ones connected to the deep channels—the ones that breathe steam when the moon waxes full. Said he was mapping forgotten trade routes, but his eyes were too hungry for history."

Bran's scales shifted with interest. "Did he mention specific locations?"

"Three names kept coming up: Red Kiln, where the basalt still hisses through cracks even centuries after the last proper firing; Temple of Nine Bellows, though half its flues collapsed after the Mantle Wars; and Emberwell Step, where tar springs feed natural flames that never fully die." The foreman shrugged massive shoulders. "Old places, mostly forgotten. Nothing there worth trading for."

As the others gathered information through words and evidence, Lysera moved through the waystation, reading emotional tremors and unspoken fears. She sat beside a woman mending clay vessels, saying little but feeling much—the undercurrent of anxiety, the whispered prayers to minor household gods for protection.

"The air feels different since he passed through," the woman said without prompting, her fingers never pausing in their work. "Like the moment before a storm, when the world holds its breath." She glanced up at Lysera, recognition flashing in her eyes at the seer's marked temples. "You feel it too, don't you? The counting. Like someone's measuring out the last moments of something precious."

Lysera nodded, a chill crawling up her spine despite the heat. "How many days since he left?"

"Three. But he said he'd return when the count reached nine." The woman pressed a small clay amulet into Lysera's palm. "Wear this. Some protections are better than none, when old things wake."

By midday, they reconvened beneath a lean-to hung with dried gourds that rattled like hollow bones in the hot breeze. The structure offered meager shade but enough privacy for their purpose. Bran spread a parchment map across their knees, the paper crackling with dryness as he smoothed its edges.

"Here," he said, marking three locations with precise dots of ink. "Red Kiln, where hot basalt vents still hiss through cracked stone—a remnant of when this region was geologically active, thanks to Pyrrhos for that. Temple of Nine Bellows, its collapsed flue buried in ash but still drawing air from deep underground channels. And here, Emberwell Step, fed by sticky tar springs that burn without human tending."

Theryn leaned over the map, his finger hovering above the markers without touching the parchment. "Three points of a triangle," he murmured. "Not random. A containment pattern."

"These kilns don't just burn," Bran continued, his voice dropping to avoid curious ears. "They breathe. They draw air from the deep places and exhale it changed. The old practitioners believed such sites were thresholds where the world grew thin, where offerings might reach hungry gods more directly."

"Or where something might pass through from the other direction," Lysera added, her fingers unconsciously touching the clay amulet now hanging at her throat.

Edrin traced the route with calloused fingertips, connecting the three points and forming a triangle, enclosing a significant portion of the region. His face was grim with determination, jaw tight with the same anger that had driven him since Emberfield burned.

"If Malrik's preparing some kind of ritual, these are his anchors," he said. "We need to break at least one point of the triangle before he completes whatever he's started."

Bran nodded, already calculating distances and travel time. "Notice: no key leylines, nothing but roads and paths. Someone chose these places specifically because they are hard to get to. I wonder if that is important? Red Kiln is the furthest but most significant—the largest and oldest of the three. That's where he'll finish his work."

"Then that's where we'll stop him," Edrin said, the statement falling between them like a hammer stroke on hot metal.

Bran nodded his assent, but then stopped nodding, "But we are going to have to hit all three. We can't just go to the Red Kiln; we have to know what he is doing at the other two sites to be able to stop whatever he's ultimately doing at that last stop." He paused for a beat and changed gears, "I wish I knew what caused him to change. When we were together that day and night, he seemed so jovial, so committed to wandering. Now this..." He couldn't finish the sentence.

"He is being led, lied to, guided by someone with much deeper intentions than just this ritual. We have to find out what he's doing to understand what the mystery woman behind him actually wants," Lysera added.

They packed their gear with the swift efficiency of those who knew that time itself had become an enemy. The map went into Bran's leather satchel, protected between sheets of stiffened hide. Around them, the waystation buzzed with midday business, travelers and traders oblivious to the pattern being traced across their lands.

Before the Solaryn's Crown sun could scorch their shoulders with its peak heat, they were moving again. Four shadows stretched long across the dusty road, heading northwest toward the first point of Malrik's triangle, each step measured against a countdown they could feel but not yet fully comprehend.

#

Umbersday, 27 Skycrest found them standing before an old abandoned Keylie Path entrance—where reality thinned like worn fabric, its usual shimmer now fractured and sickly. A recent stitch-storm had left its scars: the air buckled where it should have flowed, and what once appeared as an elegant fold in space now opened like a gaping wound. Bran pressed his chronometer against the entry stone, and the device's needle spun wildly before settling into a frantic oscillation. He withdrew the map and searched.

"This isn't on the map, it's not a recognized Path entrance, yet someone has used it. The seams are yawning open," he pronounced, scales dulling with disappointment. "The Path's integrity has been compromised beyond safe passage until it resets."

Edrin approached the threshold, hammer in hand. When he extended it toward the shimmer, the metal head trembled and emitted a low, dissonant hum that set teeth on edge. He withdrew it quickly, the surface now frosted with a crystalline layer melting almost instantly in the morning heat.

"Not an option, then," he concluded.

Lysera stood back from the rippling tear, her fingertips trembling as she counted under her breath: "Seven, eight, nine... reset. Seven, eight, nine... reset." Her eyes darted between points of empty air as if tracking invisible movements. "The voices are getting louder. Clearer. A woman is speaking through the tears in the world, and she's counting too."

Theryn moved to her side, his voice dropping to a murmur meant only for her. "Focus on the counting itself, not what's being counted. The rhythm can anchor you when the meaning threatens to drown."

She nodded, taking a steadying breath. "Seven days to bank the coals," she whispered. "We've lost three already."

They retreated from the damaged Keylie, forced onto the mundane paths that mortals had relied upon since before the Mantle Wars. By midday, they secured passage on a creaking merchant's cart pulled by two bony, sun-bleached mules. The roads were little better than the animals: rocky, rutted affairs that jarred bones and tested patience.

"We lose a day for every two we might have saved on the Keylie, but we weren't aware of it previously," Bran calculated, bracing himself against the cart's side as it lurched over a particularly vindictive stone. He unrolled his map, holding it steady despite the jolting. "We must reach the Temple of Nine Bellows by tomorrow's dawn if we're to have any hope of catching Malrik's trail before it grows cold."

Through the afternoon and evening, they traveled, stopping only when darkness made the roads treacherous. They rose before first light on Thalday, 8 Solaryn's Crown, pushing the reluctant mules to their limits, arriving at the Temple of Nine Bellows at dawn.

The temple was less a structure than a memory of one—a semicircle of scorched pillars surrounding a depression in the earth where the main chambers had once stood. Now only the foundation stones remained, blackened by ancient fires and more recent attentions. The collapsed flue protruded from the center like a broken bone, its brick throat clogged with debris yet still somehow drawing air with a soft, rhythmic sound that mimicked labored breathing.

Bran approached it cautiously, instruments in hand. "The temple still lives, after a fashion," he observed, holding a thin metal rod above the flue's mouth. The rod wavered, pulled downward by an invisible current. "It draws air into the deep places and returns it changed. The Stoneworkers would love to map this place. If only there were time to reach out to them."

Their search revealed signs of recent activity: half-buried braziers arranged in a precise arc, their contents still smoldering with unnatural persistence. Around each brazier, footprints marked with mirror-ash led away from the temple—not in a single direction, but in radiating lines like the spokes of a wheel.

"Deception," Theryn announced, kneeling beside one set of tracks. His fingers hovered above the ash without touching it. "He's laid false trails. This is decoy work."

Bran collected a sample of the ash, rubbing it between his fingers and noting its unnatural shimmer. The particles clung to his scales, their luminescence pulsing in time with the temple's breathing. "This carries Mirrathar resonance," he said. "It's not simply mirror-dust—it's been alchemically altered. Encoded, perhaps, with intention."

Edrin circled the site, hammer ready. "So which trail is real?"

"None of them," Theryn answered. "And all of them. The real path isn't marked by footprints but by the ash itself—what it carries on the wind."

They spent the day examining the temple grounds, documenting the positions of the braziers, and collecting samples from each. As night fell, Bran approached a small encampment of Sythari nomads whose scaled skin shimmered in the firelight.

Their kind were distant relatives to Bran's people, though leaner and adapted to the arid landscapes. After a brief exchange in a clicking language that made Edrin's ears itch, two of the scouts agreed to aid them.

"They've noticed the changed winds," Bran explained upon returning. "Game animals flee from certain directions without a visible threat. They believe something unnatural is occurring and will help us track it."

On the ninth morning, Bran and the two Sythari scouts—their bodies sinuous and efficient in movement—sketched wind-roses in the dirt with methodical precision. They worked silently, occasionally testing the air with long, forked tongues, capturing particles too subtle for human senses. After an hour of calculations, the elder scout dragged a clawed finger across the diagram, creating a line pointing eastward.

"The ash plume drifts against the natural currents," Bran translated. "Whatever Malrik has activated here feeds a larger working at the Red Kiln. We knew that the temple was never his final destination—merely a point in his pattern."

Theryn nodded, unsurprised. "The triangle forms, one vertex at a time. He's drawing power from each site, banking it like coals for a forge."

Throughout this work, Lysera stood apart, her lips moving in a silent tally. She had begun marking the count on her forearm with a piece of charcoal. Straight lines grouped in fives, a primitive calendar that measured something only she could perceive. Her eyes remained unfocused, seeing beyond the physical landscape to something that left her face drawn and pale.

"Nine," she said suddenly, the word cutting through the morning stillness. "Blood on the ninth. But it hasn't happened yet—the counting resets." Her fingers trembled as she made another mark on her skin. "There's a voice behind him, guiding the ritual. A woman's voice, cold as mirror-glass."

Edrin approached her, concern etched in the lines around his eyes. "What does it mean?"

"It means we still have time," she answered, her gaze clearing momentarily. "But not much. The Emberwell Step is next, and then the Red Kiln for the final working."

As they prepared to depart, the Sythari scouts offered a parting gift—a pouch of dried herbs that, when burned, would temporarily mask the group from magical detection. Bran accepted with formal gratitude, tucking the pouch into his satchel alongside his growing collection of ash samples.

"Whatever Malrik intends," he said as they set off toward Emberwell, "he's not working alone. The precision of this ritual bears the mark of someone who understands the Loom's deeper mechanics."

Edrin spat on the ground, "Flame the Loom! I don't care about that. I just don't want a god's champion running around with the power to create loom or glyph surges, destroying towns, cities, or even continents, at will!"

The path ahead shimmered with heat as the Skycrest day reached its peak. Somewhere ahead, the Gypsy Knight continued his work, each step bringing the world closer to whatever threshold he sought to cross.

#

Auronday, 29 Solaryn's Crown arrived in a copper glow, the burnt-orange horizon giving Emberwell its name. The district sprawled across a shallow depression where the land seemed to be sinking into the embrace of what locals called the Deep Pitch. Tar pockets bubbled up through cracks in the stone and sometimes ignited of their own accord. The air hung thick with the scent of mineral and decay, a primordial perfume clinging to clothing and skin like a possessive memory.

Edrin led their approach, hammer ready, eyes scanning the landscape for movement. What he found instead was stillness—a peculiar absence of the usual scavengers and carrion birds that typically haunted such places. The silence pressed against them like an invisible wall.

"There," Bran pointed toward a thin column of smoke rising from a fissure in the ground. "Tar smoke, but something's wrong."

They approached cautiously. The smoke curled upward in a perfect spiral, too deliberate in its movement to be natural. More striking was what they didn't see—no vitrified glass surrounding the source, though the heat should have transformed the sandy soil into a glassy crust.

"Unnatural fire," Lysera whispered, her eyes reflecting the smoke's coil. "It burns without consuming."

Bran knelt beside the fissure, instruments already in hand. His scales flexed and contracted as he measured the heat with a metal rod coated in color-changing salts. "The temperature is consistent with ordinary tar fires, yet nothing around it bears the marks of such heat."

Edrin circled the phenomenon, his boots leaving prints in soil that should have been baked solid. "It's as if the fire exists in two places at once—here and somewhere else."

Theryn moved with the careful precision of one accustomed to detecting falsehoods. He approached from the opposite direction, head tilted as if listening to a distant conversation. "The fire is real," he said finally. "But its purpose is deception—to draw attention while the true work happens elsewhere."

Near the smoke column stood a single wooden rod embedded upright in the mud. A chant stick with its surface carved with spiraling glyphs that matched the smoke's pattern. Four notches had been cut into its side with a blade still sharp enough to leave clean edges. When Edrin touched it, the wood radiated warmth like living flesh.

"A ritual marker," Bran explained, examining it without disturbing its position. "Four notches. Four completed phases."

"Out of seven," Lysera added, her voice distant as she counted under her breath. "He's more than halfway through the banking cycle."

They established a temporary camp nearby, rotating watches through the night.

#

At dawn on Commonsday, 30 Skycrest, they expanded their search, mapping a widening circle around the tar smoke column. The day was spent in methodical exploration, each discovery adding to the pattern they were slowly assembling.

By midday, Theryn had located a brick foundation where a structure once stood. Perhaps a way-shrine or traveler's rest. He knelt beside it, fingers hovering above and tracing patterns invisible to the others.

"There's a pillar of heat here," he said, eyes half-closed in concentration. "Not from fire, but from intention. Someone has laid a pattern of deception so thick it generates its own warmth."

He pressed his palm flat against the foundation, then withdrew it sharply as if burned. His fingers left no print on the stone, but a faint shimmer lingered where he had touched—an oily rainbow that faded even as they watched.

"Mirror-magic," he confirmed. "But perverted from its true purpose. Instead of revealing truth through reflection, it creates layers of false realities that trick even the elements." He looked up at the others, his usual enigmatic expression replaced with something harder. "This is Mirrathar's work, but corrupted. Whoever guides Malrik knows how to bend illusion into something solid enough to burn."

Thinking of who might benefit from a godlike avatar under their control, Edrin's memory immediately flashed to the Mantle Council and their meeting; his face darkened. "So, the Council woman—Vaelis— she has been a pebble in our shoe to get the council to recognize this threat. Is she behind this?"

Bran shook his head. "Not necessarily. The Mirrathar traditions are ancient, and the clan has been excommunicated, and I had thought all dead. But anyone with enough forbidden knowledge and effort can do mirror magic. And, though she is resisting our bid to get the Council to do anything, she is not even a part of the Mirrathar clan. Though our records on them, by their design, aren't very complete. There could be other practitioners."

"But few with access to both the knowledge and the materials," Theryn countered. "Whoever wields this power understands the Loom's structure intimately. They're not just pulling threads—they're reweaving them."

The implications settled over them like a physical weight. If someone was manipulating Malrik to create a ritual powerful enough to affect the Loom itself, then the stakes were far higher than a local surge or isolated disaster.

On Thalday, 1 Solaryn's Crown, beginning of the summer solstice, their understanding expanded further when a group of farmers appeared just after dawn, their faces weathered by sun and worry. They approached the camp hesitantly, led by an elderly man whose spine had been bent into a permanent question mark by decades of tending crops.

"You're tracking the fire-bringer," he stated, no question in his voice. "We've seen signs. Thought you should know."

The farmers described ash-paste hoofprints leading northeast toward the Red Kiln. Not normal horseshoes, but prints that burned the ground where they landed, leaving behind a sweet, tarry scent that made livestock refuse to graze nearby.

"Marked in sevens, they were," a younger farmer added, nervously twisting his cap. "Seven prints, then nothing for seven paces, then seven more. Like the horse was only half in our world."

Bran documented their testimony while Theryn questioned them about other signs—changes in wind patterns, unusual dreams, animals behaving strangely. The answers built a picture of subtle disruption spreading outward from Malrik's path like ripples in still water.

After the farmers departed, Lysera spoke what they were all thinking: "He's binding the ritual to the number seven. Seven days to bank the coals, seven prints to mark the path. It's not just symbolic—it's mathematical. He's creating a formula that the Loom will recognize."

"Or that a god will recognize," Edrin added grimly. "Have I said that I hate the gods?"

Theryn chuckled, "No. You have shouted it, exclaimed it, and beat us over the head with it, but you never said."

#

The morning of Lioraday, 2 Solaryn's Crown brought them to a stretch of road where the surface had partially melted and reformed into glassy shards. The aftermath of an intense heat, transforming dirt and stone, and leaving perfect patterns in its wake. Bran knelt beside the largest fragment, removing a tuning fork from his collection of instruments.

He struck the fork against his palm and held it to the glass. The note that emerged vibrated through the air with unnatural persistence, causing their weapons to hum in sympathy—spear shafts resonating, leather jerkins quivering as if alive. Bran's eyes widened as the sound took on harmonics that shouldn't have been possible from a single strike.

"Pyrrhos," he whispered, the name itself seeming to increase the vibration. "This carries the fire deity's resonance. Not just invoked, but present. Responding."

Edrin felt the hammer at his belt grow warm. "Malrik's calling on Pyrrhos? Why? What could he want from the god of fire and rebellion?"

"Power," Theryn answered simply. "The power to rewrite reality. To burn away what is and forge what might be." He looked northward, where the horizon shimmered with heat haze. "But Pyrrhos is mercurial, as likely to consume the summoner as grant their wish. Someone is taking a terrible risk."

Bran carefully wrapped the glass shard in a square of treated cloth, adding it to his growing collection of evidence. "The ritual is progressing faster than we anticipated. If Malrik has indeed secured divine attention, then the final phase—the blood offering—may come sooner than the traditional nine days."

Lysera's fingers unconsciously traced the tally marks on her arm. "The counting has changed," she confirmed. "The woman's voice is growing more insistent, pushing him to hurry before interference arrives." She looked at each of them in turn. "She knows we're following. How does she know that?"

No one responded.

They packed their minimal camp and continued northeast, following the ash-paste trail. The land began to rise, the flat plains giving way to low, rolling hills where ancient volcanic activity had left the soil rich but unstable. Ahead, the Red Kiln waited—the final vertex of Malrik's triangle, the place where whatever he had set in motion would reach its culmination.

Time, never their ally, now seemed to fold in on itself. Each hour seemed shorter than the last as they raced toward a confrontation whose shape they could only begin to guess.

#

Pyrrday, 3 Solaryn's Crown broke like a fever, the dawn pallid and weak as it crawled across a landscape stretched taut by drought. High summer had begun, and even the mornings were hot. The grasslands they traversed had forgotten their purpose; what should have been rippling fields of amber and gold had withered to brittle stalks, each broken stem releasing a dying whisper of dust. Heat rose from the ground in visible waves, distorting the horizon into a trembling mirage that promised water where only parched earth remained.

They had been walking since first light, each step measured against the urgency driving them forward and the exhaustion threatening to drag them down. The sun climbed higher, its heat merciless on their reddened skin, when Lysera spotted the first anomaly—a brief flicker of flame on a distant ridge, gone almost instantly.

"Fire," she said, voice cracked from thirst. "Northwestern ridge."

Edrin squinted against the glare. "Could be a traveler's camp."

"No," Bran countered, scales dulled to a pewter gray in the relentless light. "Wrong color. Natural flame doesn't burn violet."

They altered course slightly, angling toward the ridge. As they approached, more fires appeared—pinpricks of light dancing along the horizon like fallen stars trying to climb back to the sky. Some burned blue-white, others a deep crimson that had no place in nature's palette. Most unsettling was how they moved, not with the random flicker of true flame but with deliberate, almost choreographed precision.

Theryn studied the display, his expression darkening. "Illusions," he pronounced, the word falling between them like a stone. "Mirror witch's handiwork." He raised a hand, fingers splayed, and seemed to push against something invisible. The nearest fire wavered, stretched like pulled taffy, then vanished.

"How can you tell?" Edrin asked, hammer half-raised as if ready to strike the phantom flames.

"True fire casts shadows outward," Theryn explained, eyes never leaving the ridge. "These cast shadows inward, feeding on themselves. It's a signature of mirror-magic—light bent back upon itself until it believes it's something it's not." His mouth twisted in distaste. "She's either trying to mislead us or mark us for something else's attention."

"Or both," Lysera whispered.

They continued forward, the phantom watchfires multiplying as the day stretched toward evening. Some appeared close enough to warm skin, yet left no heat impression. Others materialized directly in their path, forcing detours that added precious time to their journey. By sunset, the entire horizon seemed ablaze with false light, a corona of illusion that enclosed them in a shrinking perimeter.

Theryn moved ahead of the group, dismissing the more insistent phantoms with gestures that looked half mystic, half irritated. "She's mapping our progress," he called back. "Each illusion we interact with tells her where we are, how fast we're moving, how many we number."

That night's rest brought little respite. They maintained vigil according to the season names. Theryn took the first watch, Greenthread, Lysera took Sunweft, Bran claimed Loomreach, and Edrin took the deepest hours of Mooncall. They slept in brittle shifts, each dream interrupted by the quiet movements of their companions or the distant howl of something that might have been wind.

#

Dawn of Umbersday, 4 Solaryn's Crown found them already walking, bodies stiff from too little rest on unyielding ground. The phantom fires had retreated with the night, but new torments replaced them. Dust devils spun up from nowhere, carrying laughter-like whispers and heat mirages that painted the horizon with impossible cities and lakes.

Edrin's jaw tightened with each mile. The muscles beneath his skin clenched visibly with determination or suppressed rage—perhaps both. The hammer at his belt seemed heavier, its metal head absorbing the sun's punishment and radiating it back against his thigh. Yet he set the pace, each stride purposeful, his eyes fixed on the smudge of darker land marking their destination.

Lysera walked beside him, her fingers curled around the edge of her cloak, knuckles white with tension. The blue veins at her temples pulsed more prominently now, pausing mid-step, and listening to words carried on frequencies only she could detect. Each time, she would resume walking with a small shudder, her lips moving in silent tallies.

Behind them, Bran consulted his almanac with scholarly diligence, tapping his staff against each milestone they passed.

The instrument was both a walking aid and a measuring tool, notched at precise intervals to calculate distances when pressed against landmarks.

"We maintain our pace," he announced after one such calculation, voice clipped with the effort of speaking through parched lips. "If the ritual follows traditional patterns, we should reach the Red Kiln before the final phase begins."

Theryn ranged ahead and to the sides, his gaze constantly scanning every horizon. The heat haze did not seem to affect his vision; where the others saw wavering distortions, he perceived clear lines and true shapes. Every false marker—illusions subtler than the watchfires, mere suggestions of shade or water—he identified and warned the others away from.

"She's clever," he admitted, returning from one such scouting foray. "The deceptions grow more convincing as we near our goal. Not just visual now—I've found patches where the air itself tells lies about distance and direction."

By midday, the grasslands gave way to a rockier terrain marked by black glass and twisted volcanic stone formations. The air carried the mineral tang of hot springs and the faint, sweet rot of sulfur. Ahead, the land rose in a series of stepped ridges, reminding Edrin of the forge ladders in Emberfield. Each tier was a challenge, and each summit revealed only another climb.

They paused in the narrow shade of a stone outcropping, rationing water from canteens grown tepid in the heat. Lysera pressed her palms against the rock, seeking coolness that the stone refused to provide.

"The counting accelerates," she said suddenly, eyes unfocused. "She pushes him to complete the ritual before the proper time. The pattern wants nine, but she demands seven."

Bran looked up sharply. "That would disrupt the entire resonance structure. If the blood phase begins before the full banking cycle completes—"

"Then whatever's meant to come through may arrive incomplete. Damaged." Theryn finished the thought, his expression grim. "Or hungrier for what it's missing."

Edrin pushed away from the outcropping, hammer already in hand. "Then we keep moving. Reach the Red Kiln before sunset. End this before it begins."

They shouldered their gear and resumed their march, four figures dwindling against the vastness of the parched landscape. The brittle wind carried the scent of ash as they pushed onward. Each step measured against a countdown they could feel thrumming in the earth beneath their feet.

Ahead, though still beyond visual range, the Red Kiln waited—final vertex of Malrik's triangle, focus point of whatever transformation he sought to bring into the world. And with each mile they traveled, the phantom watchfires reappeared on the ridges above, marking their progress with cold, unblinking vigilance.

#

Dusk draped the land in ambiguous mercy—relief from the sun's tyranny but harbinger of a darkness that concealed as much as it revealed. The party camped beneath a twisted thornbush with branches splayed against the darkening sky. Exhaustion had become their fifth companion, dragging at their limbs and clouding their judgment. Even Edrin, with his forge-hardened resolve, conceded that a few hours' rest might prevent a fatal misstep tomorrow.

They built a fire small enough to warm but not betray. Its flames huddled close to the earth as if sharing in their caution. Bran made a thin broth from their meager supplies—barely sustenance, yet enough to restore what the day's march had drained. They ate in silence, each lost in private calculations of distance and time, of what might await them at journey's end.

Sleep came in shallow waves, washing over them and retreating, never deep enough to truly restore. Theryn took first watch, his silhouette motionless against the starlit horizon as he scanned for illusions and more tangible threats. Lysera curled on her side, one hand pressed to the earth as if trying to read the land's intentions through her fingertips. Edrin and Bran slept back-to-back, the hammer between them like a shared talisman.

The night had just touched its deepest point when Lysera bolted upright, her cheeks pale as moonlight, eyes wide and unseeing. Her breath came in short, panicked bursts, and when she spoke, the words emerged as if dragged from some interior depth where nightmares bred.

"Breath on the seventh has passed," she whispered, the sound carrying unnaturally in the still air. "Blood on the ninth comes next."

The others woke instantly, hands reaching for weapons, eyes searching the darkness for threats. Lysera's gaze fixed on something beyond the physical landscape, her pupils dilated to black pools.

"The ritual accelerates," she continued, voice steadying but no less urgent. "She pushes him to complete it before the proper interval. The pattern demands precision, but her impatience outweighs tradition."

Bran was already packing, his movements swift and economical. "Who is she, Lysera? The Mirrathar woman? The mirror witch?"

Lysera's focus returned slowly, like someone emerging from deep water. "The voice behind his counting. Cold as glass, hungry as fire." She pressed her palms to her temples. "I see the brick ring burning. Not with natural flame—with something that eats light instead of creating it. He kneels at its center, carving glyphs into his own flesh that pulse in time with the earth's breathing."

Edrin extinguished the fire with a handful of dirt, his movements betraying no hesitation. "We move now," he said, the decision already made. "How far to the Red Kiln?"

"Three hours at forced march," Bran answered, shouldering his pack. "Less if we take the direct route across the glass fields."

They broke camp. Lysera's words hung in the air like smoke, a prophecy propelling them toward the distant brick ring she saw in her vision. Even the night seemed to part before their urgency. The darkness thinned as they climbed higher, and the stars lent what meager light they could.

By dawn, they had reached the outer perimeter of what locals called the Glass Fields. A stretch of land where ancient volcanic activity had liquefied the soil and cooled it in a way that mimicked a frozen sea. Here, they found the first concrete evidence of Malrik's passage: wagon ruts fused into black glass, the tracks melted and reformed by heat far greater than anything nature produced.

Bran knelt to examine them, pressing his palm against the surface. "Still warm," he confirmed, scales contracting against the unexpected heat. "Whatever passed through here did so recently, and it burned hotter than forge-fire."

Edrin crouched beside him, his own calloused hand confirming the lingering warmth. The glass was smooth as polished metal yet structured like frozen waves, each ripple capturing the moment of its transformation from earth to something otherworldly. Within the glass, mirror-flecks glinted like fallen stars, catching the rising sun and throwing back tiny rainbows that danced across their faces.

"Mirrathar dust," Theryn observed, his expression grim. "Mixed with something else—something that binds it to physical matter instead of letting it remain ethereal." He pointed to a line of flecks that formed a subtle pattern within the glass. "She's not just helping him—she's embedding her own work inside his. A ritual within a ritual."

They followed the glass-fused tracks higher, the land rising in a series of natural terraces, sculpted further by whatever force Malrik had unleashed. With each step, the air grew thinner but also charged with a prickling energy.

As they crested the final ridge, the Red Kiln's brick ring loomed through a wavering column of heat. It was older and larger than they had imagined—a perfect circle of rust-colored bricks, some thirty paces across, set into the hilltop. At its center, a depression filled with what appeared to be smoldering coals sent lazy spirals of crimson smoke into the morning sky.

The entire structure pulsed with a dull red glow, coming from within the bricks themselves, as if the kiln were a living organ of the earth exposed to air. Around its circumference, smaller brick formations created a labyrinthine approach, forcing any visitor to spiral inward, passing through seven consecutive thresholds before reaching the central chamber.

Edrin drew his hammer, its weight familiar and reassuring in his hand. His eyes locked on the shrine's smoking heart, jaw set with the same determination that had carried him from Emberfield's ruins.

"We have one day left," he said, voice low and resolute. "Tomorrow, he spills blood."

Bran nodded, unfastening his chronometer to check its reading against the sun's position. "If Lysera's vision holds, he will attempt the final phase at dawn tomorrow—Auronday. A convergence of calendar and deity that amplifies the working. We are still in the Sunweft festival season, the gods are close and listening."

"Then we stop him before dawn," Edrin replied, already calculating approaches, angles of attack, contingencies. "Whatever he's trying to bring through stays on the other side."

They stood outlined against the sunrise, weapons ready, faces hardened by days of pursuit and nights of uneasy rest. Below them, the Red Kiln waited, ancient and patient, indifferent to the dramas of mortals yet integral to their resolution. Within its brick embrace, the final act of Malrik's ritual would either be completed or broken. And with it, perhaps the fate of more than just those present.

Lysera's eyes drifted to the sky, where the last stars faded against the strengthening blue. "The counting never stops," she murmured. "Even when we sleep, she counts. Even when we fight, she counts. Seven, eight, nine… reset."

Theryn laid a hand on her shoulder, steadying. "Then we break the count," he said simply. "And see what happens when the rhythm fails."

They descended toward the Red Kiln, their shadows stretching long across the glass fields, four figures moving against a landscape transformed by powers they had only begun to understand.

#

Dawn on Auronday, 6 Solaryn's Crown, birthed itself in shades of crimson and gold that seemed to bruise the sky. The road leading to the Red Kiln had transformed overnight. What had been merely glass-fused wagon ruts now spread across the entire path, black waves frozen in mid-surge, as if the earth itself had attempted to flee whatever power had reshaped it. Heat rose from the stone under Edrin's palm, steady and unnatural, as if the hill itself breathed in measured counts.

The air carried a sweet, tar-resin tang, coating the back of their throats, metallic and cloying. It tasted of intention, as if someone had distilled ambition itself and set it aflame. Near the path's edge, a charred prayer strip fluttered from a thornbush, its fabric blackened but still intact. A sun glyph had been heat-scarred into a sliver of mirror affixed to the strip's center. The symbol warped by flame still was potent enough to make the light around it bend and distort.

Bran approached the prayer strip cautiously, tuning fork in hand. He struck the instrument against his palm and held it close to the mirrored glyph. The fork hummed a low note resonating through their bones, a bass vibration stirring memories none of them possessed—ancient fires, chaotic rebellions, the exhilarating freedom of destruction turned toward renewal.

"Pyrrhos," Bran confirmed, his scales paling as the tuning fork continued to vibrate long after it should have fallen silent. "The fire deity has answered someone's call. This isn't merely ritual invocation—it's active communion."

Lysera stood slightly apart, her eyes distant and unfocused as she tracked movements beyond physical sight. The veins at her temples pulsed visibly, blue channels that carried visions instead of blood.

"Seven nights banked," she whispered, voice layering into multiple timbres as if more than one person spoke through her lips. "Blood next. Her voice... counts with him. She's eager, impatient. The pattern demands nine, but she pushes for seven." She blinked rapidly, trying to clear her sight. "She knows we're here."

Above them, the Red Kiln's brick ring glowed dull crimson in the morning light, the color deepening to something almost purple at its center. Smoke curled upward in seven distinct spirals, merging before dissipating against the dawn. The structure pulsed like a heart exposed to air, each throb sending tremors through the ground that matched the rhythm of Lysera's counting.

They approached with weapons drawn, moving in formation practiced during their journey. Edrin at the lead, hammer raised; Bran to his right, staff prepared to inscribe counter-glyphs; Lysera behind them, her Sight providing warning of unseen threats; Theryn circling to flank, his movements liquid and unpredictable.

The Red Kiln's interior revealed itself as they crested the final approach. At its center knelt Malrik, his red bandana now stained black with soot and something darker that might have been blood. He had shed his vest and shirt, exposing a torso covered in self-carved glyphs pulsing with unnatural light—some fresh enough that blood still seeped from their edges, others scarred over as if carved days or weeks earlier.

Before him stood a makeshift altar assembled from broken wagon parts and seven distinct stones. Each vibrated at a different frequency that combined into a dissonant chord. On the altar, a flame burned without fuel that consumed nothing, its light folding inward. Malrik's hands moved through the flame without burning, fingers tracing patterns in the air that left momentary scars in the fabric of perception.

He looked up as they approached, his expression shifting from deep concentration to something like relief.

"You're late," he called, his voice retaining its characteristic showman's projection despite the gravity of the moment. "I half-expected you days ago. The counting gets lonely when it's just me and her."

Edrin advanced, hammer held ready. "Step away from the altar, Malrik. Whatever you're trying to call through, it ends now."

Malrik laughed, the sound hollow and echoing strangely from the kiln's walls. "Call through? You misunderstand. I'm not calling anything—I'm offering. Myself, primarily." He gestured to the glyphs carved into his chest and arms. "Seven sigils of preparation, two of binding, one of transformation. The pattern is ancient, but the application..." He smiled, teeth white against soot-blackened skin. "That's all new."

Bran began inscribing a counter-glyph in the dust at his feet, staff moving with precise, urgent strokes. "He's attempting to mantle," the lizardfolk hissed. "To become a vessel for divine power."

"Not just any power," Malrik corrected, rising to his feet with fluid grace that belied his carved and bleeding state. "Pyrrhos. The ember that ignites revolution. The fire that cleanses what is corrupt and forges what might be." His eyes flickered with an internal light that matched the glyphs on his skin. "The Loom is failing. You've seen it—surges, collapses, reality fraying at the edges. The gods withdraw, or sleep, or simply cease to care. Someone must wake them."

"By becoming their puppet?" Edrin challenged, taking another step forward.

"By becoming their champion!" Malrik's voice rose, carrying an edge of fervor that hadn't been present in their previous encounter. "Their hands in this world. Their will made flesh."

Before Edrin could respond, the flame on the altar surged upward, reaching toward the open sky. Within its center, something moved—a presence neither human nor wholly other, a consciousness that seemed to regard them with curious disdain.

"Enough talk," Theryn called from the kiln's perimeter, where he'd positioned himself behind Malrik. "The ritual is nearly complete. Break it now or face what comes after."

The battle erupted without further preamble. Edrin charged, hammer swinging in an arc that should have connected with Malrik's shoulder. Instead, the Gypsy Knight pivoted with impossible speed, the blow passing through empty air. The hammer struck the altar instead, shattering one of the seven stones. The discordant chord faltered, momentarily weakening the flame pillar.

Malrik retaliated with a gesture that sent a whip of fire lashing toward Edrin's face. The blacksmith ducked, feeling heat scorch his hair as it passed overhead. Bran completed his counter-glyph, his hand resting on a glyph on his travel clothes, and activated it with a sharp command word. The symbol flared blue-white, sending a wave of nullifying energy across the kiln's interior that temporarily dampened the glow of Malrik's carved sigils.

Theryn moved like a shadow given form, flanking Malrik and striking at the man's blind spots with precise, economical movements. His attacks never aimed to kill—only to disrupt the ritual's rhythm, to break the pattern of seven that Malrik had so carefully constructed.

Throughout the chaotic exchange, Lysera remained at the kiln's edge, her voice rising in counterpoint to the battle's percussion. She called warnings when the flame pillar pulsed with renewed strength, directed Edrin away from false targets, and guided Bran toward weaknesses in the ritual structure.

"The third stone from the left," she shouted. "It's the lynchpin!"

Edrin pivoted mid-strike, redirecting his hammer toward the indicated stone. It shattered with a sound like a struck bell, releasing a wave of pressure that knocked all combatants backward. The flame pillar wavered, thinning to a thread.

Malrik's expression shifted from confidence to panic, then to a strange, listening stillness. His head tilted as if hearing instructions from a voice only he could perceive. Then, with a sudden decision, he slashed his palm with a mirror-shard dagger and pressed the bleeding hand against the largest of his carved glyphs.

"Blood on the seventh!" he shouted. "Not the ninth. The seventh!"

The flame pillar collapsed inward, folding into Malrik's chest where his blood met the glyph. His eyes flared with inner fire, and for a moment, his face seemed to overlay with another's—a woman's features, cold and calculating, there and gone so quickly it might have been imagination.

Before they could close the distance, Malrik turned and fled toward its far wall. He passed through what appeared to be solid brick as if it were nothing more than mist, leaving behind only a ripple in the air to mark his passage.

The battle ended as abruptly as it had begun. The altar's remaining stones cooled to ordinary rock. The ember-fire died without fuel to sustain it. Only the carved glyphs on the kiln's floor remained, their outlines still faintly glowing with residual power.

Bran approached these markings cautiously, instruments already in hand. He knelt to examine what appeared to be mirror-shard sigils singed into the stone—not simply scratched or painted, but burned into the very molecular structure of the brick. The patterns were complex: spirals within spirals, lines that turned at precise angles to create shapes that hurt the eye if followed too closely.

"Mirrathar illusion-work again," he confirmed grimly, running a scaled finger just above the surface without touching it. "These are too perfect, too mathematically precise for Malrik to have constructed on his own." He looked up at the others. "Someone with intimate knowledge of the Loom's structure is using Malrik as a conduit."

"The mystery woman," Edrin said.

Theryn moved to stand beside Bran, studying the sigils with narrowed eyes. "The mirror-magic's touch is unmistakable. Someone is playing a deeper game than mere manipulation." He glanced at the spot where Malrik had vanished. "Whoever she is, she didn't just guide his ritual—she's embedded herself within it. The voice Lysera hears is counting alongside his."

Edrin paced the kiln's perimeter, hammer still ready, searching for any sign of Malrik's return. "We stopped him this time, but he'll try again. The ritual was incomplete."

"But not unsuccessful," Bran cautioned. "He achieved something. The fire deity responded, even if the full transformation didn't occur. Next time, he'll be stronger, more prepared. Though

I can't be sure, it seems like there are two Malriks, the one who is loud and boisterous, and the other, the calculating, partially mantled god playtoy who doesn't even know that someone is using him. He is not responsible for what he did here today. *She*, whoever she is…" He tailed off as Lysera moved closer.

Lysera approached the center, her steps careful, navigating around the sigils. "The counting continues," she said softly. "Different now—not seven, eight, nine, but a new sequence. Three, five, thirteen. A Loom sequence, I see, ancient and rarely used."

Theryn sighed, rolling his shoulders as if shedding a weight. "I normally avoid combat when possible," he said, his usual enigmatic smile returning. "But some patterns must be broken before they can be rewoven. This one…" He gestured to the sigils. "This required intervention."

He turned to face them fully, something shifting in his posture that suggested imminent departure. "I must follow a different thread now. This mirror magic leaves traces I can track better alone." He pulled one of his roll-ups and lit it, puffing on the smoke.

Before anyone could object, he stepped backward into a shadow cast by the kiln's wall—and simply vanished, the darkness folding around him like a cloak. One moment present, the next gone, leaving only the faint impression of his smile lingering like the after-image of a bright light, and a faint scent of sweet smoke.

Edrin stared at the space, then at Bran. "Does he always do that?"

"Since I've known him," Bran nodded, already packing away his instruments. "He appears when needed, vanishes when done. He'll return if our paths converge again." Bran flicked his tongue out and sniffed the air, smiling at the faint scent of smoke that he had come to associate with Theryn.

The three remaining companions gathered at the kiln's center, surveying the aftermath of their confrontation. The battle was won, but the larger conflict had only just revealed itself. Malrik was merely a vessel; someone was the true architect. And whatever design this hidden woman had set in motion continued to unfold, counting down toward some culmination they could only begin to guess.

Edrin hefted his hammer, studying the mirror-shard sigils one last time. "We follow him," he said, the decision already made. "Find out what he's planning next. Stop it before it starts."

Bran nodded, calculating. "The ritual pattern suggests he'll need to recover before attempting another working. Three weaves, perhaps more."

"No, we have three wefts," Lysera said, her voice stronger now, more certain. "Three threads or less to prepare for whatever comes next."

They left the Red Kiln as the sun reached its zenith, the brick structure already cooling. Behind them, the mirror-shard sigils continued to pulse with faint light—a beacon, perhaps, or a warning of what still waited to be born from fire and reflection.

—CHAPTER SEVEN—

Temple *of the* Earthflame

Chapter Seven

The Temple of the Earthflame – Dawn (Morday, 19 Solaryn's Crown)

The Temple of the Earthflame rose before them like the spine of some ancient beast half-buried in earth—vertebrae of weathered stone reaching toward a cloudless, merciless sky. Wind had scoured the outer walls to resemble parched skin, while the inner chambers lay partially collapsed, their stones tumbled inward as if the structure itself had sighed and surrendered to gravity's persistent pull. What stood defiant: archways carved with fire glyphs catching the light at certain angles, throwing back patterns that the eye could not quite follow without discomfort. The flue that gave the temple its name lay buried beneath decades of ash and dust. Yet somehow, it still drew breath—a subtle inhalation that could be felt rather than heard, maintaining communion with whatever lay beneath.

Edrin led the way, his boots crunching over fragments of tile with elaborate mosaics depicting the ascension of Pyrrhos from molten depths. His hammer hung heavy at his side, its head still bearing scorch marks from their confrontation at the Red Kiln. Behind him, Bran moved with scholarly precision, pausing occasionally to note the temple's unusual dimensions in his ever-present almanac. Lysera walked with her arms wrapped around herself, as if the wind carried voices only she could hear.

Near what must have once been the altar, a small fire burned with surprising vigor given the temple's exposed state. A woman knelt before it, her back to the visitors, hands extended toward the flames. When she rose and turned to face them, the sunlight caught on her fingers, flashing with prismatic brilliance. Each digit was adorned with rings of blackened silver, and set within each ring was a sliver of mirror no wider than a hair, yet positioned with such precision that they transformed her hands into constellations of captured light.

"Pilgrims seek the ember; the flame receives them," she intoned, her voice carrying the dry crackle of someone who spoke more often to fire than to people. "I am Cantor Yselle Keth, keeper of this threshold. What brings you to the Temple of the Earthflame when its glory has long since cooled?"

She was neither young nor old—one of those ageless devotees whose features had been baked into permanent serenity by proximity to their chosen deity. Her robes were rust-colored, embroidered with spiraling flame patterns mimicking those carved into the temple walls, and her hair was pulled back into a severe knot pierced by a ritual pin made from volcanic glass.

Bran stepped forward, scales catching the firelight. "We track a pattern across the land. A ritual that leaves glass where earth should be, and calls on powers that should remain dormant."

Edrin cut in, less patient with formalities. "A man passed through here. Calls himself the Gypsy Knight. Marked with self-carved glyphs, attempting to mantle Pyrrhos."

Yselle's expression remained placid, but something flickered behind her eyes—recognition, certainly. Her ringed fingers traced a pattern in the air between them that left momentary afterimages, as the trail of a torch swung through darkness.

"The ember seeks a vessel; the vessel seeks the flame," she said, the cadence rhythmic and practiced. "Yes, a pilgrim passed this way. Not like the others who come with small prayers and smaller offerings. His devotion burned with… unusual clarity."

Lysera stepped closer, her eyes fixed on Yselle's rings. "Did he speak of counting? Of sevens and nines?"

"The sacred intervals," Yselle nodded. "Seven days to bank the coals, two to prepare the vessel, one for transformation. He knew the ancient rites, though his urgency suggested… improvisation." She tilted her head, studying their faces with renewed interest. "You interrupted his work at the Red Kiln. The flames whispered of your interference."

"The flames whisper many things," Bran observed diplomatically. "Including, perhaps, where he might have gone next?"

Yselle turned toward her small fire, passing her ringed hands through it without apparent discomfort. The mirror slivers caught the light, throwing fractured rainbows across the broken altar stones. For a moment, it seemed she might not answer, lost in communion with her deity. Then her shoulders straightened, decision made.

"Emberwell Step," she said finally. "Where the earth has a mouth that breathes true flame. He spoke of completing what was started, of finding a threshold where the veil is thinner." She gestured toward the distant horizon with a sweep of glittering fingers. "Three days' journey southeast, where tar springs feed eternal fires. The Step remembers what stone forgets."

Edrin exchanged a glance with Bran and Lysera. The time frame aligned with Bran's calculations—three days for Malrik to recover from the partial mantling, three days for them to intercept him before his next attempt.

"He won't succeed," Edrin said, the words half promise, half threat.

Yselle's lips curved in a small smile, which carried neither warmth nor malice. "What comes through a doorway depends on who opens it, and what waits beyond. The Gypsy Knight may think he calls Pyrrhos, but it is not always the summoned who answer."

"Pilgrim trinkets," came a voice from behind one of the broken columns, the sound carrying a playful edge. "Careful where you point those—they catch more than just firelight."

Theryn materialized from shadow as if he'd been woven from it, his movements fluid and precise. He hadn't been there a moment before—or at least, had not been visible—yet now he stood smiling, gesturing toward Yselle's mirror-sliver rings.

The cantor stiffened, her composure briefly fracturing into something sharper and more guarded. Her hand instinctively covered the rings, fingers curling inward as if to shield them from his scrutiny.

"The Riddleweaver arrives uninvited," she said, her voice now pitched low with warning. "This is consecrated ground."

"All ground is consecrated to something," Theryn replied, his smile never faltering. "The question is what claims the consecration, and whether it speaks with its own voice." He circled toward them, never quite approaching directly, his path a loose spiral that both honored and mocked the temple's ritual patterns. "Those pretty little mirrors—they reflect more than they reveal, don't they? I wonder whose eyes look back from them when you're alone."

Yselle's face hardened into the rigid mask of someone accustomed to defending her faith against skeptics. "The fire purifies. The mirror clarifies. Only those who fear truth would question the tools of revelation."

"I question tools wielded by those who may not understand what they grasp," Theryn countered, his voice gentle despite the challenge in his words. "Not all reflections are faithful to their source."

The tension stretched between them like a thread pulled taut enough to sing. Then, before Yselle could respond, Theryn simply… wasn't there. No flash, no dramatic exit—just a whisper of displaced air, a few swirling dust motes, and the fading impression of his enigmatic smile. Gone as abruptly as he had appeared, leaving only questions in his wake.

The cantor exhaled slowly, her ringed fingers resuming their rhythmic passing through the flames. "The Riddleweaver has always sown doubt where faith should flourish," she said, more to herself than to her remaining visitors. "But doubt burns as readily as certainty when fed to the right fire."

Edrin watched the space where Theryn had been, his expression caught between frustration and grudging admiration. "That little trick of his could get flaming tiresome."

"Him appearing and disappearing like smoke?" Bran shrugged, already calculating their next move. "It's part of his nature. He follows threads we cannot see, and rejoins us when it best suits his needs. But he is helpful," He added weakly, "sometimes. But yes, it can, and does, get tiresome."

Lysera touched Edrin's arm, her focus already shifting toward their next destination. "Emberwell Step awaits. Three days' journey, but we can make it in two if we push." Edrin looked at her hand and felt a jolt. Lysera misread his look and hurriedly pulled her hands back, her face worried that she had upset him.

As they took their leave of the temple, Yselle called after them, her voice carrying on the wind that never ceased its scouring. "When you find him, remember: it is easier to break a vessel than to contain what spills from it."

The Temple receded behind them, its half-collapsed walls framing a sky that deepened in color with each step they took. Above them, clouds began to gather, their edges tinged with the faintest hint of flame.

Verdantia – Morning (Morday, 20 Solaryn's Crown)

Verdantia's market square unfurled beneath a quilt of mismatched awnings, each shade competing with its neighbor in a silent debate over which color best repelled the midday heat. Merchants arranged their wares like offerings to fickle gods of commerce—pottery glazed in blues deep enough to drown in, spices heaped in earthen pyramids that perfumed the air with promises of distant lands, fabrics that whispered when the breeze caught them. The cobblestones beneath it all had witnessed centuries of such displays, worn smooth by generations of haggling people, their surfaces retaining the morning's coolness even as the sun climbed higher. Somewhere in the market's heart, a blind musician coaxed melodies from a stringed instrument, the notes floating above the hum of commerce like birds riding thermal currents.

Alira Marrowind occupied a modest corner where morning shade lingered longest, her small wooden stall a testament to practical simplicity rather than mercantile ambition. Her fingers worked with practiced efficiency as she arranged jars of healing salves in neat rows, each container with a label identifying its purpose: burn relief, fever reduction, bone-knitting, sleep inducement. The glass caught the light differently depending on each salve's contents—some cloudy as river stones, others clear as tears, a few shimmering with suspended particles.

Unlike the vendors who called their wares with theatrical flourish, Alira waited in patient silence. Those who needed her remedies would find her; those who didn't would pass by, untroubled and untroublesome. Her Fairykin heritage manifested in subtle ways—the warm, earthy tone of her skin glimmering faintly in direct sunlight, the deep auburn of her hair threaded with hints of green and gold, the forest-green eyes that held a grounded clarity rather than ethereal shimmer. Her attire was practical yet distinctly her own: an earth-toned tunic adorned with embroidered flowers and vines telling stories in their winding patterns.

The hooded figure approached her stall as someone seeking specific medicine, but with the measured steps of someone wishing to avoid drawing attention. Beneath the hood, Alira glimpsed only the curve of a clean-shaven jaw and lips pressed into a line of concentration. The visitor's hands emerged from voluminous sleeves—pale, unmarked by labor, with nails trimmed to perfect crescents.

"Your remedies are known for their potency," the figure said, voice pitched low enough that it might have been any gender. "I seek something for clarity of thought."

Alira nodded, reaching for a small jar filled with amber liquid. "Three drops under the tongue before sleep. The mind finds its center when the body rests."

The hooded visitor made a show of examining the jar, turning it to catch the light. "An excellent craft," they murmured, then leaned closer, voice dropping further. "But I bring something requiring a different kind of clarity."

Their hands moved with the swift precision of a practiced magician. One moment, they held the jar; the next, they pressed something small and cold into Alira's palm and placed three silver coins on the counter. The transaction happened so quickly that anyone watching might have seen only a customer paying for medicine.

Alira's fingers closed around the object instinctively. The moment it touched her skin, something shifted beneath her feet—the cobblestones rippled like black glass for a heartbeat. The sensation was so brief she might have imagined it, yet the coldness against her palm remained very real.

"The elders will know what to do with it," the visitor whispered, already withdrawing. "Veilwood waits for your return."

Before Alira could respond, the hooded figure melted back into the market crowd with the same deliberate anonymity with which they had arrived. She glanced down at her hand, opening her fingers just enough to glimpse what she'd been given: a cameo brooch, small enough to conceal in a closed fist, its surface etched with what appeared to be a woman's profile rendered in Mirrathar silver-blue. A hairline crack ran across the profile's temple, barely visible except where it caught the light.

She slipped the cameo into an inner pocket sewn specifically for items that required discretion. The weight of it seemed disproportionate to its size, as if it carried more than just physical substance. Her mind already turned toward Veilwood and what the elders might make of such an artifact, but her journey would have to wait until the market's end. Drawing attention through a hasty departure would serve no one.

As Alira rearranged her jars to fill the space left by the sold remedy, she became aware of a conversation at the neighboring stall, where two women vendors had their heads bent over similar objects spread on a velvet cloth.

"These new pieces flooding the market are clever forgeries, but they're not genuine Mirrathar work," the older woman insisted, adjusting a headscarf that had slipped during her emphatic gesturing. Her fingers were stained with decades of handling dyes and metal polishes. "The genuine ones always have that hairline crack across the profile—it's not a flaw but a signature. Like the master craftsman is reminding us that even perfection contains the seed of its own undoing." The old woman grimaced, "A very astute clan, those ancient Mirrathar."

The younger vendor snorted, holding a jeweler's loupe to her eye, examining one of the cameos. Her face bore the pinched expression of someone who had learned to distrust through experience rather than wisdom. "The crack is easy enough to fake," she countered. "What they never get right are the eyes. Counterfeits make them too flat, no depth. You can craft the prettiest face in the world, but if the eyes don't look back at you, it's just dead silver."

She lowered the loupe, her verdict apparently confirmed. "Besides, real Mirrathar work leaves a trace you can feel, not just see. My grandmother had a true piece—it made the hairs on your arm stand up when you held it. Said it was because the Mirrathar embed part of themselves in every creation. Not just craft, but intention."

Alira turned away from her jars, pretending to search for something beneath her stall while straining to hear more. The cameo in her pocket seemed suddenly heavier, as if responding to being the subject of discussion.

"That's exactly it," the older woman agreed, lowering her voice to a near-whisper. "The real ones change things around them—just a little, just for a moment. My brother swears he saw one turn water to glass for the space of a breath. It's not just metalwork; it's mirror-magic set into physical form."

"Mirror-magic," the younger one scoffed, though her tone carried a hint of unease. "Next, you'll tell me the Loom responds to pretty jewelry."

"Mock if you want," came the tart reply, "but there's a reason the Mirrathar keep to themselves. What they create isn't meant for common hands, or common understanding."

Alira carefully repacked her unsold remedies, her movements deliberately casual despite the quickening of her pulse. The cameo's weight against her side seemed to pulse in rhythm with her heart, as if awakening to the knowledge being spoken about it. Whatever the hooded figure had given her, it carried significance beyond its craftsmanship.

The Veilwood elders would need to see it immediately. Some things could not wait for convenience; some messages required prompt delivery. As the market day wound toward its close, Alira mentally plotted the fastest route back to her people, her fingers occasionally brushing the hidden pocket where the Mirrathar cameo rested—a small piece of someone else's intention, now entrusted to her care.

Emberwell Step – Twilight (Commonsday, 21 Solaryn's Crown)

Emberwell Step revealed itself to them as dusk gathered the world's colors and compressed them into deeper, more solemn hues. The plateau rose from the surrounding flatlands like a titan's abandoned altar, its edges worn smooth by millennia of wind and the occasional summer storm. Across its surface, dozens of fissures exhaled plumes of tar smoke—thin, spectral columns twisting skyward before dissipating against the darkening canopy. Each breath of smoke carried the mineral tang of earth's innards, a scent both primal and patient, as if the land itself were sighing out secrets too old for human comprehension. The wind caught these offerings and scattered them, revealing that what should have been a site of true flame instead presented only the memory of fire, a theater of heat without its central actor.

Edrin crested the rise first, hammer already loosened in its harness. The party had pushed themselves mercilessly since leaving the Temple of the Earthflame, stopping only when darkness made the roads impassable or when Lysera's counting reached such fevered intensity that she could no longer walk straight. Now, seeing what awaited them, Edrin's face hardened into lines of disappointment mingled with suspicion.

"No glass," he muttered, scanning the ground where the smoke emerged. "Not a single patch."

Bran moved to his side, instruments already in hand. The lizardman's scales had dulled during their journey, the usual metallic sheen reduced to the flat luster of pewter left too long unpolished. Still, his movements retained their scholarly precision as he knelt beside the nearest fissure and extended a slender rod over the smoke.

"Temperature inconsistent with true flame," he confirmed, examining the rod's color-changing tip. "This is residual heat only—earth-warmth, not combustion." He rose, turning in a slow circle to take in the entire plateau. "These are decoys. Designed to mimic the appearance of the Step, but carrying none of its power."

Lysera hung back, her arms wrapped around herself. Her lips moved in that now-familiar silent tally, fingers twitching against her sleeves as she tracked some rhythm only she could perceive. The blue veins at her temples stood out starkly against skin gone pale with exhaustion and concentration.

"Seven, eight, nine… reset," she whispered, just loud enough for the others to hear. "He's completing the seventh cycle. The ninth approaches."

They spread out across the plateau, each searching for something that might contradict the growing certainty that they had been misdirected. The smoke columns offered no answers, merely continued their slow, hypnotic dance against the deepening twilight. Stars emerged overhead, cold points of light watching the proceedings with ancient disinterest.

Bran discovered the shrine tucked between two larger fissures—a crude arrangement of stones piled into a rough pyramid, crowned with a half-melted wagon strut. Before it stood a single wooden rod embedded upright in the earth, notched with precise cuts catching the last of the day's light. He crouched before it, nose twitching as he processed the scents.

"Another chant stick," he called to the others. "It's been used to mark ritual progression."

Edrin and Lysera converged on his position, picking their way carefully between the smoke plumes. The stick stood approximately waist-high, its surface darkened as if by fire or strong sunlight. Four distinct notches had been carved into it at equal intervals, each cut clean-edged and deliberate.

Bran carefully extracted a thin blade from his collection of instruments and scraped residue from within one of the notches. The material was black but carried an oily sheen that caught the light oddly, refracting it in directions that defied simple physics. He brought the sample to his nose first, then, with the caution of long experience, touched the tip of his tongue to a minuscule portion.

"Tar, certainly, as a base," he said, his voice taking on the detached quality that emerged when analysis supplanted emotion. "Merryvine, which traditional texts prescribe for memory and purification." He paused, considering the aftertaste. "And something else—a sweetness that doesn't belong. Binding oil, I think, though altered from its standard composition."

Lysera's eyes flashed with recognition. "Binding oil is used to yoke intentions to physical objects. To make concepts tangible."

"Or to bind one ritual to another," Bran added, carefully preserving the remaining sample in a small glass vial. "This could be an anchor point, tethering whatever is happening here to workings elsewhere."

Edrin had moved beyond the shrine, examining the ground with methodical thoroughness. His boots left clear impressions in the soft soil, but he found no evidence of the glass-fused footprints they had come to associate with Malrik's passage. No true heat signatures, no vitrification where flame should have transformed the surrounding stone.

"Another false trail," he concluded, returning to the others with shoulders tight with frustration. "Just like the watchfires that led us on a chase on the road to the Red Kiln. Mirages, wasting our time while he completes his work elsewhere."

The disappointment was palpable—three days of forced marches, limited rest, and dwindling supplies, only to find themselves following shadows and smoke. But Bran's expression had shifted from disappointment to a more calculating assessment as he studied the chant stick.

"Not entirely a waste," he said, rising to his feet with newfound urgency. "The fourth notch confirms we're on the right timeline—just the wrong location." He pointed to the notches, tracing their progression. "One for each completed phase of the ritual. Four notches means he's halfway through the banking cycle."

"And the seventh approaches," Lysera added, her voice distant as she picked up the count again. "Seven to bank, nine for blood."

Bran nodded, already repacking his instruments with swift efficiency. "The Red Kiln," he said, the words falling between them with the weight of certainty. "That's where the true rite will culminate. We were correct the first time."

Edrin's jaw tightened as he processed the implications. "You're saying we've been running in the wrong direction while he circles back to where we started?"

"A deliberate misdirection," Bran confirmed. "These decoys were never meant to convince us completely—only to delay us until the ritual progressed beyond our ability to interrupt it."

Lysera's eyes cleared momentarily from their visionary distance. "The woman's voice grows louder," she said. "More insistent. She's trying to hurry him, pushing the intervals closer together."

"Then we have even less time than we thought," Edrin concluded, already turning back toward the path they had climbed to reach the plateau. "How long to return to the Red Kiln?"

Bran performed swift calculations, fingers tracing invisible notations in the air. "Four days at normal pace. Three if we push beyond reasonable limits."

"Then we push," Edrin said, the decision already made. "If this is truly the fourth phase, we have until the seventh to intercept him. After that..."

He didn't need to finish. They all understood the stakes: after the seventh phase came the ninth—blood on the altar, the threshold crossed, something invited through that should remain separate from their world.

They gathered their gear with grim efficiency, knowing their survival meant abandoning the false path. The smoke columns continued their slow dance, indifferent to the departure of their audience. As the three figures descended from the plateau, the first stars emerged in force, their cold light merely a witness to the urgency that drove tired limbs onward.

Behind them, Emberwell Step faded into the gathering darkness, its deception complete yet ultimately unsuccessful. Whatever game Malrik and his unseen puppet-master played, the rules were becoming clearer—a ritual split across multiple sites, anchored by binding oil and mirror-magic, racing toward a culmination that would forever alter the balance between mortal and divine.

The night stretched before them, vast and unknowable. Somewhere within its embrace, the true countdown continued, a rhythm they could feel thrumming beneath their feet with every step they took.

Outside of Emberwell Step – Afternoon (Morday, 20 Solaryn's Crown)

Edrin squinted at the late afternoon sun dipping low over the rugged trail ahead. By his reckoning, Red Kiln was still a few days away beyond those dusky hills, and a heavy dread coiled in his gut. He pulled the reins, slowing his tired horse as Bran and Lysera caught up. The silence among them was tense. Even the wind seemed to hold its breath, as if reluctant to break the news none of them wanted to voice.

At last, Edrin exhaled a hard breath. "We're not going to make it in time," he said quietly. His voice was flat, but anger and worry flickered in his eyes. "Even with these fresh horses, we are going to be late."

Lysera closed her eyes as if pained but nodded. "Malrik's ritual…" she began, her tone hushed. "If he finishes it before we get there—"

"*When* he finishes it," Theryn, who wasn't there one minute and suddenly was, on a horse, right behind Edrin, interjected bitterly.

Edrin, reacting purely on nerves, turned around with his hammer in his hand, ready to strike. "How do you do that?" Edrin exclaimed, "And on a horse no less."

Theryn smiled his little smile. He hopped down from his saddle and began pacing, running a hand through his hair. "He's probably lighting the damned kiln as we speak. That snake got a head start." Theryn tried for a wry grin to mask his frustration, but it fell flat.

Bran slid off his horse and crouched by the roadside. He picked up a handful of red-dusted earth and let it trickle through his fingers. "Smoke," he murmured, nostrils flaring. "Can you smell it?" Everyone paused. On the breeze was a faint trace of something acrid—perhaps a distant fire or the foretaste of Malrik's impending conflagration. Bran's keen sense of smell had caught what the others hadn't. "Malrik's doing or not, something's burning far off. And it's too far for us to reach by tonight, or even tomorrow."

Edrin cursed under his breath and swung down from his horse, boots hitting the ground hard. "So what do we do? We can't just give up." He unrolled a crumpled map against his thigh, scanning their route with a grim frown. A tense silence followed, broken only by the rustle of the map and the distant caw of a crow circling overhead.

Theryn kicked a pebble off the road, eyes narrowed. "We ride faster. Through the night without rest?" he offered, though his tone was doubtful. They all knew it wouldn't bridge a multi-day gap.

Lysera wrapped her arms around herself. She looked exhausted—dark crescents under her eyes, her usually calm face drawn with strain. Still, her voice was steady. "Even if we did, our horses would collapse. And Malrik… he'll complete the ritual by tomorrow's dawn. That's what the auguries indicated." Her words hung in the air like a death sentence.

Edrin's jaw tightened. He remembered all too well their last confrontation with the Threadmere Council, a mere week ago. The elders in their marble hall had listened to the warning about Malrik and the Red Kiln with polite skepticism and ultimately withheld aid. Edrin had left that meeting seething. The thought of returning to Threadmere empty-handed made his pride flare. Yet here they were, out of time and options. He exchanged a look with Lysera, seeing his own apprehension mirrored in her face. "The council…" he started reluctantly, "we could go back. Appeal to them again."

Theryn let out a short, harsh laugh. "Really? Go back to those cold-blooded scholars and beg *again*? They practically shoved us out the gates last time." He mimed a grand bow with a sarcastic flourish. "Thank you, wise council, may we have another 'maybe later' while the world burns?"

Lysera shot Theryn a reproachful glance. "They didn't outright refuse," she insisted softly. "They were afraid of interfering without proof. Maybe now, with so little time, they'll listen. We have to make them understand what's at stake." There was a desperate edge to her plea. She pressed a hand to her temple briefly, as if staving off a headache. "Perhaps the Loom's threads have shifted enough to change their hearts. Even Vaelis may yet change her mind."

Edrin ruminated, folding the map slowly. Threadmere was most of a day's and night's ride back west. If they pushed hard, they could be there by morning. It was a gamble. Malrik's ritual clock was ticking, but confronting Malrik alone and late was a guaranteed failure. At least in Threadmere, they might find resources or glyph scholars to help. He hated the idea of detouring, but Lysera was right: they had to try something.

Meanwhile, Bran stood abruptly, eyes alight with an idea. "Threadmere… ley lines… Wait." He brushed the red dust from his hands and stepped closer to the group. "There might be another way to reach Red Kiln fast."

Theryn quirked an eyebrow. "Unless you're also part Aerithari, Bran, I'm not seeing it." He stopped for a second. "Wait, you aren't, are you?"

Bran shook his head. "Not flying—Keylie Paths." Seeing the blank looks from Theryn and Edrin, he explained, words tumbling out in excitement. "Not leyline travel. There's an old hub—or at least it was *rumored* to be—near Red Kiln. If we could find someone in Threadmere to send us through the Path…"

Edrin's brows furrowed. "Keylie travel is dangerous and highly restricted. You need a trained witch for that. Do you really think Threadmere has one just waiting for us?"

Bran's eyes gleamed with curiosity and hope. "Threadmere is the heart of glyph academia. If any city has leyline scholars to open a Keylie Path, it's there. And I swear I heard a tale in the taverns about a Keylie hub by the Red Kiln—leftover from some old war or maybe a guild outpost. If there is a hub which still functions, we could… we could effectively *jump* most of the distance."

Lysera's tense face softened a fraction as she considered Bran's proposal. "I've heard of the Keylie Paths in old hymns," she said. "Threads of the world that allow instant travel for those who know the weaving." Her gaze sharpened with renewed energy. "It could work. It bypasses space… and it might be our only hope to arrive before Malrik unleashes whatever horror he's prepared."

Theryn, ever cautious, folded his arms. "Or it might spit us out into a mountain or the middle of the ocean. These are *ancient* magics. Who even knows how to navigate them now?" Despite his skepticism, a spark of interest glinted in his eye. The fact that Theryn hadn't dismissed the idea outright meant he, too, saw it as the slimmest of lifelines. He thought for a moment, "You know. The fey used to have special powers to navigate the leylines. We could take a trip to their realm." Then he shook his head, immediately discounting that idea, "Ah, that would take longer and require more skills than we poor folk have now. At least *you* poor folk. I might be able to reach them on my own."

"Threadmere's council will know," Edrin said, sounding more convinced as he spoke. He straightened his shoulders, resolve hardening. "We go back to Threadmere. We convince the council to help—either with soldiers, or a leyline witch, or whatever it takes. If there's truly a Keylie Path near Red Kiln, we'll make use of it. Somehow."

Bran was staring at nothing, his mouth moving with some sort of unvocalized chant, "Wait, didn't we see a Keylie marker the first time that we traveled to the Red Kiln? It looked abandoned and disused." He deftly pulled the map he had been filling out as they traveled. "Yes, there it is. It was still days away from the Red Kiln. If that marker is still there, there might be a hub or marker closer. We didn't have time to search around the Red Kiln."

Theryn let out a long breath, nodding slowly. "Alright. It's a long shot… but long shots are all we've got left." He managed a half-smile directed at Bran. "And here I thought you just had your nose in the air for fun. Good sniffing out that bit of lore, Bran."

Bran chuckled, relieved. "I try." He glanced eastward, toward Red Kiln, and the smile faded from his face as that faint smell of distant smoke lingered. "Hang on a little longer, Red Kiln. We're coming."

Lysera was already adjusting her horse's bridle, anxious to move. "I'll gather what light we have left," she said under her breath—one of her subtle invocations. The setting sun's rays seemed to cling a moment longer on the horizon as if heeding her request.

Edrin looked over each of his companions, pride and concern mingling in his expression. And when he glanced at Lysera, he noticed, again, how powerful she looked, yet so fragile. He felt a tug at his thoughts and heart. What was it about her?

They were exhausted, on edge, and now about to rush back the way they came. But they were united in purpose. "Threadmere, then," he said, voice firm. "Let's ride. We've no time to waste."

With that, the four mounted up again and wheeled around, hooves pounding against the dirt road back west. As the last light of day bled out in red streaks across the sky, a sense of urgency drove them onward. None of them noticed the crow take flight from its perch and glide east toward Red Kiln, its dark shape vanishing into the coming night. Each was focused on the new plan—fragile and desperate as it was—hoping the council in Threadmere would hear their plea, and that somewhere in ancient stone beneath that city lay a doorway of magic still waiting to be used. It was a gamble against time and fate, but for the first time since morning, a flicker of hope warmed their path.

Chapter Eight

Threadmere

Chapter Eight

Threadmere – Morning (Auronday, 21 Solaryn's Crown, TT 1215)

The Council chamber unfurled its ancient grandeur like a living tapestry woven from stone and light. Morning sunlight poured through stained-glass windows depicting the Mantling Conclaves of ages past, casting pools of azure, emerald, and amber across the white marble floor. The chamber had witnessed centuries of debate, decision, and destiny—its high ceiling collecting whispers and proclamations alike, its stone benches worn smooth by generations of robed councilors who had gathered to thread the needle between divine will and mortal necessity. Today, the councilors filed in with grave steps; the chamber seemed to hold its breath, sensing that the patterns about to be proposed might forever alter the weave of their world.

Timepieces ticked from the walls—seven of them, each marking a different divine interval, their hands moving in perfect synchrony despite their varied faces. The metallic hum of machinery countered the rustle of robes and muted greetings as the councilors took their seats. Archive Keepers occupied the lowest tier, their ink-stained fingers clutching scrolls and almanacs. Above them sat representatives from the merchant guilds and craft halls, their attire more ostentatious but no less formal. The highest tier housed the elder councilors—those who had survived enough political winters to earn their elevated vantage.

At the center of the chamber, a circular table of polished obsidian reflected both light and shadow in equal measure. Twelve high-backed chairs surrounded it, eleven already occupied by the Council's innermost circle. The twelfth chair—larger than the others, its back carved with intertwined threads forming the sigil of the Loom itself—awaited the Council Leader, who now entered from a side chamber. Silence rippled outward from her entrance, conversations faltering and then ceasing altogether as she took her place.

Among the eleven at the high table sat Councilor Vaelis, her posture almost seemed artificial, as if she were a sculpture brought to reluctant life. Her dark blue robes, embroidered with silver thread, caught the light with every subtle movement, emphasizing the pallor of her skin. Her face was a study in controlled beauty—features balanced with such precision they might have been drawn by a master artist whose hand never trembled. Yet it was her eyes that commanded attention, not for their unremarkable gray color, but for the intensity burning behind them.

Around her neck hung a pendant: a jagged shard of mirror wrapped in delicate silver filigree. Her gloved fingers—always gloved, regardless of season or setting—occasionally brushed against this pendant with a gesture that might have seemed unconscious were it not so precisely timed.

When the Council Leader nodded in her direction, the veiled adept rose with fluid grace, her movements calculated to draw the eye without seeming to seek attention. She placed her palms flat against the obsidian table, her reflection gazing back at her with a fidelity that would have unnerved anyone who truly understood what they were seeing.

"Esteemed colleagues," she began, her voice carrying throughout the chamber without apparent effort. "We gather in troubled times. The Loom-surges increase in frequency and intensity. Our borders face incursions from those who sense our vulnerability. Even the weather betrays us, with droughts where there should be bounty, floods where there should be gentle rain." She paused, allowing the weight of these observations to settle upon the assembly.

The chamber remained silent, councilors leaning forward on their benches, some concerned, others with the careful blankness of those who preferred to reveal nothing until they had measured which way the political winds might blow.

"We have tried the old ways," the Councilor continued, her gloved fingers tapping almost imperceptibly against her pendant. "We have distributed divine blessing like crumbs scattered for birds, hoping that enough might fall where needed most. Yet the crumbs grow fewer, and the need grows greater." Here, her voice lowered, forcing those present to strain to hear, drawing them in like fish to a lure. "The gods' reserves are finite. This is a truth we have long known but seldom acknowledged."

Murmurs rippled through the chamber at this statement—some affirmative, others uneasy. To speak of limits to divine power was to approach heresy, yet none could deny the evidence of waning influence. Loom-surges had damaged temples, ley lines had shifted without warning, and prayers that once yielded consistent results now went unanswered with alarming frequency.

Vaelis allowed the murmurs to rise and then fall before she continued, her timing impeccable. "I propose a solution both ancient and new. Rather than dispersing divine power in inefficient dribbles across countless petitioners, we should select a single mortal champion—a vessel worthy to receive concentrated divine blessing, capable of wielding it where most needed."

The fairykin advisor seated near the chamber's eastern wall straightened, the delicate structures of her wings catching the colored light. Elsewhere, a guild representative with hands roughened by decades of stonework leaned toward his neighbor, whispering calculations of cost and benefit. At the Archive Keeper's bench, Bran Silvertail's scales shifted hue almost imperceptibly, the only outward sign of his internal disquiet.

"We must channel their power efficiently," Vaelis declared, sweeping her gaze across the assembly, making momentary eye contact with key figures whose support she had already cultivated through private conversations and carefully placed favors. "Through one worthy vessel who can direct divine intervention with precision rather than scattering it with hope and prayer."

Several councilors nodded in agreement, particularly those from regions recently affected by Loom-surges. Others exchanged uncertain glances, their expressions caught between tradition and desperation. The council had always advocated for balanced distribution of divine favor—it was the very foundation of the Concord Weave Edict that had ended the Mantle Wars. Yet the evidence of that system's failure had become increasingly difficult to ignore.

After the practitioner's proposal, the chamber fell still for three breaths—just long enough for ideas to root or die. Then came the whisper of scales against fabric as Bran Silvertail rose from his seat among the Archive Keepers. Light caught on his pewter-hued scales, now shifting toward a deeper bronze as blood warmed beneath them—a subtle tell known only to those familiar with lizardfolk physiology, betraying the intensity of his focus despite his measured movements. His clawed fingers retrieved a worn almanac from his belt, the book's binding cracked and mended so many times that it resembled a palimpsest of repairs—history preserving history.

Bran's Archive Keeper's robes rustled softly as he straightened to his full height. Unlike some of his kin who hunched to appear less imposing among humans, Bran stood with the straight-backed dignity of one who had earned his position through decades of meticulous scholarship rather than political maneuvering. His eyes, glowingly softly purple, with their distinctive vertical pupils, surveyed the chamber briefly before settling on Councilor Vaelis with a detached curiosity.

"Honored Council," he began, his voice soft yet carrying a weight that belied its volume. "I must respectfully challenge Councilor Vaelis's proposal on historical grounds." His clawed finger traced down a page of the almanac with precision that spoke of thousands of hours spent navigating such texts. "The Archive records document seventeen prior attempts at power concentration of the kind being proposed today." He paused, allowing the number to register. "Fourteen ended in calamity. Two in partial success, followed by eventual corruption. Only one achieved its intended purpose, and even that was temporary."

A murmur passed through the chamber like wind through autumn leaves. The Council Leader leaned forward slightly, her silver circlet catching the light. "Elaborate, Archive Keeper Silvertail."

Bran nodded, turning a page with deliberate care. "In Threadturn 327, the Conclave of Eastern Provinces selected a champion to combat drought. Emmara of the High Plains received concentrated blessings from three regional deities." His voice remained neutral, but his scales darkened further as he continued. "For three seasons, rains followed her footsteps. On the fourth, the power calcified in her bones. She became rooted to the earth, unable to move, a living conduit for rain that never ceased until floods claimed seven townships."

He turned another page, claws making a soft click against the parchment. "Threadturn 512: Kevan Stoneheart was empowered to end the Border Wars. Five gods granted him strength and invulnerability." Bran looked up, meeting the eyes of several military representatives. "He ended the war, yes, by subjugating both sides under his own rule. When the gods attempted to withdraw their blessing, he had absorbed enough power to resist. It took a coalition of nine regional champions to defeat him."

The Council chamber had grown still now. The earlier murmurs were silenced as Bran continued his methodical recitation. Even the timepieces on the walls seemed to tick more softly, as if reluctant to interrupt.

"Most recently, Threadturn 899: The Mantle Council itself appointed Lysander of the Vale as vessel during the Ash Blight. Three months of miraculous healing followed." Bran's tone softened almost imperceptibly. "Then madness. Lysander claimed to hear the voices of gods fighting within his mind. He burned seven temples before taking his own life, believing it the only way to silence the divine cacophony."

Bran closed the almanac with gentle reverence, though he kept one claw between the pages to mark his place. "These are not isolated incidents or failures of individual character. They represent a pattern inherent to the nature of divine power when concentrated beyond mortal capacity to contain."

His gaze swept across the chamber, taking in the faces of those who had nodded along with the Vaelis's proposal moments earlier. Some now wore expressions of doubt; others seemed almost relieved to have reason for hesitation. Only a few maintained their earlier certainty, their eyes flickering toward the mage as if seeking counterarguments.

"The Loom's design," Bran continued, "distributes power across many threads, not through a single strand. This is not accidental or inefficient—it is essential architecture." His scaled hand made a weaving gesture, fingers interlacing to demonstrate. "Multiple threads bearing moderate tension can support great weight. A single thread bearing the same burden will eventually snap, often catastrophically."

At the high table, Vaelis' expression remained composed, her posture unchanged. A faint tightening of her gloved fingers on the pendant was her only response to Bran's systematic dismantling of her proposal. Her eyes tracked him with the intensity of a predator whose prey has unexpectedly grown spines.

From near the eastern wall, the fairykin advisor rose, her translucent wings catching fragments of colored light from the stained-glass windows. "Archive Keeper," she began, her voice musical yet pointed, "your historical perspective is valued. However, might current circumstances warrant an exception? The frequency of Loom-surges exceeds any recorded precedent. Traditional methods clearly prove insufficient." Her wings fluttered once, betraying agitation beneath her formal demeanor. "Is the risk of a champion not preferable to the certainty of continued deterioration?"

Several councilors nodded at this, the pendulum of opinion swinging again. Fear was a powerful counterweight to caution, and recent months had provided ample fuel for fear.

Bran acknowledged the question with a patient inclination of his head. "A fair inquiry, Advisor." He reopened his almanac, turning to a different section with practiced ease. "The Archive anticipated such reasoning. Following each catastrophe, similar arguments were made—that exceptional times demanded exceptional measures." His finger traced a line of text. "Loom-scholar Verithal wrote, 'The desperate solution appears most reasonable to the desperate mind, yet desperation itself blinds us to alternatives less obvious but more sustainable.'"

He closed the book again, this time completely. "I do not dismiss our present danger, nor do I suggest inaction. But historical precedent counsels against this specific action. There are other paths—more complex, perhaps requiring greater patience and cooperation—but with considerably less risk of catastrophic failure."

At the high table, the Vaelis' fingers curled around her pendant. Her face remained a perfect mask of thoughtful consideration, but beneath that façade, calculation churned. She had anticipated opposition, of course—had even prepared counterarguments for the specific historical examples Bran had cited. Yet something in the lizardfolk's presentation had shifted the chamber's energy in a way she hadn't expected. Not outright rejection of her proposal, but a dampening of enthusiasm, a planting of doubt where she had hoped to sow certainty.

"The Archive Keeper's caution is noted," she said, her voice modulated to convey respect rather than frustration. "Yet I wonder if we might distinguish between past failures of method and the method itself. Perhaps the flaw lies not in the concentration of power, but in the selection of vessels unprepared for such responsibility."

Her words hung in the air, neither fully refuting Bran's argument nor abandoning her position. The seed remained planted, merely driven deeper, where it might take root beyond the reach of historical warning.

The measured exchange of scholarly debate gave way to something closer to arguments when Edrin Ashwood pushed himself away from the gallery wall where he had been leaning. His movement carried none of the ceremonial grace permeating Council proceedings—it was the straightforward advance of a man accustomed to dealing with immediate problems rather than theoretical futures. The simple traveling clothes he wore, still bearing the dust of the road, stood in stark contrast to the formal attire surrounding him. Each step he took toward the speaking floor echoed against marble with unvarnished purpose, drawing every eye in the chamber toward this unexpected interjection.

At his side hung his father's blacksmith hammer, not as an ornament or symbol, but as a tool worn smooth by genuine use.

The wooden handle bore the imprints of generations of Ashwood hands—dark with oil and sweat, polished by countless grips in moments of both creation and desperation. The metal head, though clean, carried the marks of repeated heating and cooling, of striking and shaping, a history written in microscopic stress lines and subtle discolorations that only another smith might fully read.

"Councilors," Edrin began, his voice carrying the burr of Emberfield's regional accent, rounder vowels, and the consonants harder than the refined speech that typically filled these halls. With each word, the accent became more pronounced, as if proximity to power stripped away the veneer of formality he might have adopted for the capital. "I've come back a long way to tell you what happens when power gathers too tightly in one place. And you sit and ARGUE?" His voice rose sharply.

Several councilors shifted in their seats, uncertain whether to call for proper procedure or allow this irregular testimony to continue. The Council Leader decided for them with a single, subtle gesture—fingers spread in permission, though her expression remained guarded.

"I'm Edrin Ashwood of Emberfield, you know this," he continued, hands resting at his sides, not attempting to disguise or embellish who he was. "Or what's left of Emberfield, after the surge that took half our homes and ALL of our people."

The chamber stilled, even the shuffling of papers ceasing. Loom-surges were abstractions to many in the capital, concerning reports rather than lived experiences. Edrin's presence made the abstract suddenly, uncomfortably concrete.

"It wasn't as dramatic as you might think," Edrin said, his gaze moving from face to face, missing nothing. "No great flash of light or divine proclamation. Just the sound of reality tearing—like cloth ripping, but slower. Then the fire, the heat. Then the screaming." He paused, not for effect but because memory required its due. "My father died holding back a wall of warped stone so children could escape the schoolhouse. My brother trying to bank the fire-glyph forge before it consumes the eastern quarter. All of them dead."

He gestured toward Vaelis, though his eyes never quite met hers directly. Something in her composed stillness repelled his attention.

"You talk about champions and vessels and concentrated power," he continued, "as if it's something new. But you've seen this before. Not the fancy version with divine blessings and ceremonial mantles. The crude version, where power gathers in one place and burns too hot for anything around it to survive."

His hand found the hammer's haft, fingers curling around it with the unconscious familiarity of a gesture performed thousands of times. "After the surge, when no champions came, no vessels of divine power arrived to save us, you know what happened? Ordinary people tried to rebuild. Neighbor helping neighbor. Sharing what little remained because that's what humans do when not distracted by grand schemes. Now, I hear, constantly threatened by hordes of bandits. Even the catfolk have abandoned the borders that they used to watch over."

The Council Leader leaned forward slightly. "Your point is taken, Ashwood of Emberfield. But surely you understand the difference between destructive surges and directed divine intervention through a worthy vessel?"

Edrin's laugh held no mirth, just the rough edge of a truth too sharp to swallow. "Worthy vessels. As if worthiness protects against corruption, or power can tell the difference between ambition and duty." He shook his head. "Faith belongs to the people—not to rulers or champions or vessels. It's meant to be carried in parts, shared in community, not concentrated until it burns a hole through whoever holds it."

His words echoed longer than they should have in the vaulted chamber. Several of the elder councilors exchanged glances weighted with memory—they were old enough to recall the aftermath of previous experiments with concentrated power, though young enough then to have been merely witnesses rather than decision-makers.

"I stand before you to refuse," Edrin stated firmly, his voice finding a deeper register, resonating with the hammer at his side. "If this is some roundabout way of offering a Mantle to me or those like me who've survived surges, I refuse it. I won't carry what should be shared by many. I won't be the vessel that proves, one more time, that mortals aren't meant to channel divine power like rivers through a single gorge. Flame the gods!"

His grip tightened on the hammer's haft until his knuckles paled. "Though I despise it," he spat on the pristine floor. Every eye in the room stared at the spittle on the floor. "Some force made the Loom with many threads for a reason. Who are we to unpick their work and claim we can do better?"

Silence followed, heavy and expectant. Several councilors shifted uncomfortably, their formal robes suddenly seeming constrictive. Others appeared moved by Edrin's blunt sincerity, their expressions softening from political calculation to something more thoughtful. A fairykin advisor's wings fluttered once, then stilled, as if she had been about to rise but reconsidered.

At the high table, Vaelis maintained her perfect composure, though something flickered behind her eyes—not quite anger, not quite respect, but a recalibration of strategy. The pendant at her throat caught the light oddly, reflecting it inward rather than outward before returning to normal.

The Council Leader rose slowly, her silver circlet glinting as she surveyed the chamber. "We have heard compelling arguments both for and against this proposal. As is our tradition when faced with matters of such weight, we shall now take counsel with ourselves." She extended her hands, palms upward. "Those in favor of pursuing Councilor Vaelis's proposal for a concentrated vessel of divine power, make your position known."

Right hands, signifying agreement, rose throughout the chamber, tentative at first, then with greater conviction. Yet even as they appeared, others raised their left hands in counterpoint, then voices from both sides, breaking protocol to argue their objections rather than waiting for the formal opposition count. The neat proceeding fractured into pockets of debate, councilors turning to their neighbors with urgent whispers that quickly grew to audible disagreements.

The Council Leader allowed this for precisely seven breaths—the traditional span permitted for spontaneous deliberation—before striking a small bronze gong beside her seat. The sound rippled through the chamber, bringing immediate though reluctant silence.

"The Council is divided," she announced, her tone carrying no surprise at this outcome. "When unity cannot be achieved, wisdom dictates patience. This matter shall be deferred until our next gathering, allowing all present to consult both records and conscience." Her gaze swept the chamber, settling momentarily on Edrin before moving to Vaelis. "Neither acceptance nor rejection is rendered today."

With these words, she struck the gong again, signaling the session's end. Councilors rose in a rustle of fabric and murmured conversations. Some gathered in small groups to continue their discussions, and others departed alone with thoughtful expressions. Edrin stood his ground for a moment longer, then stepped back toward the gallery, his purpose fulfilled if not his hope.

Councilor Vaelis watched it all, her gloved fingers resting lightly on the table's obsidian surface. The Council had not embraced her proposal with the enthusiasm she had hoped for, but neither had they rejected it outright. The seed was planted. Indecision and fear would water it. And patience had always been among her more useful weapons.

As the chamber emptied, her gaze found Edrin Ashwood's retreating form, studying him with the cold calculation of one who recognized both an obstacle and a potential tool. The hammer at his side had shaped metal; her will would shape destiny. It was merely a question of applying the right pressure at precisely the right moment.

The session had adjourned with tension hanging in the air, the matter unresolved. But beneath the apparent stalemate, currents shifted and eddied like invisible threads being slowly, inexorably rearranged.

#

The grand doors of the Threadmere Council Hall creaked open as Edrin, Bran, Theryn, and Lysera stepped out into the cloudy midmorning light. The Council of Elders had been cooler than a Coldmere winter, yet notably, they hadn't denied the group's plea outright. That half-victory felt hollow. Edrin paused at the top of the marble steps, knuckles white where he gripped the strap of his satchel. "At least they listened this time," he muttered, trying to mask his frustration.

"Listened, yes," Theryn said, trailing a step behind. "And told us, 'We will consider your request with due care.'" He imitated the prim voice of Elder Serafina almost perfectly, drawing a reluctant smirk from Bran. Theryn shook his head and sighed. "Due care will get us due nothing. They'll deliberate until Malrik's smoke is on the horizon, or at their very gates."

Lysera descended the steps quietly, her long sleeves brushing the carved stone railing. "No… I think they'll act," she said, though her tone betrayed uncertainty. "They're afraid, even if they won't show it. Did you see Councilor Renald's eyes when I described the red lightning around Malrik's ritual circle? He *knows* what that could mean." She bit her lip, worry evident on her face.

Edrin frowned. "Fear alone won't make them move faster. We might not get a decision until it's too late. The sun will set in a few hours… Malrik's almost at his moment." His voice dropped so others passing by on the street below wouldn't overhear. Threadmere's white spires and bridged terraces spread out before them, bustling with scholars, petitioners, and citizens, all oblivious to the doom gathering in the east. It made Edrin's skin prickle with urgency.

Bran inhaled deeply, drawing in the familiar scents of Threadmere—old parchment, candlewax, and the tang of ozone from countless glyph spells in the air. But something new cut through the usual city smells. Bran tilted his head, brow furrowing. "Do you smell that? Metal and... vinegar?" he murmured to the others under his breath. Before anyone could respond, a figure emerged from the shadow of a nearby colonnade.

"Excuse me!" called the young woman, stepping forward. She wore a deep blue traveling cloak clasped with a plain bronze pin, and a satchel bulging with scrolls hung at her side. Alira looked barely out of her youth—perhaps eighteen threadturns—but there was a certain mature composure in her stance. She inclined her head respectfully. "I—I couldn't help overhearing in the hall. You're the ones petitioning about Red Kiln and Malrik, aren't you?" Her voice was low and reserved, but carried an undercurrent of urgency.

Immediately, Edrin and Theryn moved subtly into defensive positions. Edrin's hand rested near the haft of his hammer, Theryn half a step in front of Lysera. Threadmere might be a city of peace, but they'd made enemies enough to be cautious. "Who's asking?" Edrin replied guardedly.

Alira lifted her palms in a placating gesture. Up close, Bran noticed her eyes were a pale gray, almost silver, and they flicked nervously between the four companions. "As I said, my name is Alira, Alira Marrowind. I'm an adept from the Academy here." She took a breath. "I was in the council chamber earlier, delivering something to the elders, when I heard your appeal."

Lysera stepped forward slightly, studying the young woman. "We did ask for anyone skilled in leyline magic," she said, voice careful. "The council made no mention of an adept among them."

A wry smile ghosted over Alira's lips. "They wouldn't. I'm not exactly on the council—just an apprentice scholar, really. But I do have *some* training with the Keylie Paths." She glanced around to be sure no one was eavesdropping, then added in a softer tone, "Enough to send you where you need to go, I believe."

Bran's heart skipped. He traded a quick, wide-eyed look with Edrin. Could this be the help they'd dreamed of finding here? But Edrin remained cautious, expression unreadable.

Theryn, however, arched an eyebrow and said what Edrin would not. "That's a generous offer from a stranger. Why would you stick your neck out for us?" His tone was not unkind, but plainly suspicious. "For all you know, we're chasing fairy tales about Malrik's ritual. And from what I know of fairy tales, sometimes, they can kill those who help."

Alira hesitated, then reached to unclasp the bronze pin at her cloak. "Because I revealed this to the council today," she said. The item in her palm was a brooch: once decorative, now cracked down the center. The metal around the fracture was blackened, as if burned, and the red gemstone at its heart was cloudy with a dark swirl. A subtle hiss of energy emanated from it. "It's *mirror-tainted,*" Alira explained. "A cursed brooch I recovered from a traveler who fled the Red Kiln area. The corruption in it… It's the same kind of magic that's spreading in Red Kiln right now. I'm sure of it."

Lysera drew a sharp breath. She didn't touch the brooch, but her fingers twitched as if she could feel its aura from a distance. "I sense it," she whispered. Indeed, up close, the brooch's gemstone looked like a miniature storm of shadows within glass. "Mirror magic… Malrik may be delving into mirror realms or using reflective glyphs." Her intuition gave weight to Alira's claim.

Edrin relaxed his stance slightly, eyes fixed on the brooch. "You brought this to the council as proof? And they still wouldn't act immediately?" His voice hardened in quiet anger.

Alira grimaced. "They will, in their own time. They're not entirely refusing you. But bureaucracy can be slow, even in crisis." She carefully refastened the cloak pin, hiding the brooch once more. "I'm not content to wait while that… *monster* prepares to tear a hole in the Loom or whatever he intends. If Malrik's ritual succeeds, the mirror-curse in that brooch could spread, or worse."

Bran's curiosity burned in his eyes as he peered at Alira. "You've seen what's happening out there? What exactly is Malrik doing?"

She shook her head. "I only have pieces of the puzzle. The traveler who gave me the brooch spoke of a red haze over the Kiln and people going mad from reflections in their own drinking water. Malrik's name came up as some kind of prophet or leader. That's when I hurried here to warn the council. Then I heard your plea." Alira drew herself up, and though nervous, she looked resolute. "We have a common goal: stop Malrik and whatever he's unleashed. I can't leave Threadmere for long, but I can help you get to Red Kiln *fast*—faster than any horse or airship."

Bran all but bounced on his heels. "The Keylie Paths," he breathed. "You can send us through the leylines, can't you?"

Alira gave a single firm nod. "Yes. There's a Keylie Path hub beneath Threadmere—an old one that the council rarely uses. I found it during my studies. With the right preparation, I can open a path from there to the convergence point, if it is operational, near Red Kiln."

Theryn let out a low whistle. "You just happen to have a secret magic gateway under the city, and you're offering to let us use it. Just like that." He still looked distrustful, but a hint of hope crept into his face despite himself.

Edrin glanced at his companions, measuring their reactions. Lysera's eyes shone with cautious optimism; Bran could hardly contain his excitement; Theryn's lips were pressed thin, but he hadn't raised any objection that wasn't reasonable. Edrin turned back to Alira. "If what you say is true… if you can really do this, then we accept your help. We don't exactly have other options." He stepped closer and extended a hand to her. "Thank you, Alira. For stepping forward when no one else would."

Alira visibly exhaled, relief softening her features. She reached out and clasped Edrin's forearm in the smith's greeting. "Don't thank me yet. The Paths aren't easy or safe. But I promise I'll do everything I can to see you through."

Bran's earlier spark of suspicion had not entirely faded. As Alira turned to lead them towards the academy district, he hung back a moment and tugged on Theryn's sleeve. In a hushed whisper, he said, "This seems almost too convenient. A mirror-cursed trinket, a ready-made leyline guide… What if—"

"—She's the one behind it?" Theryn finished under his breath, eyes flicking to Alira's slight form moving ahead of them. "I've been thinking it too. She could be influencing Malrik from afar for all we know."

Lysera overheard and shook her head subtly as they walked. "She just showed a powerful dark object to the council, openly. Does that really scream 'evil mastermind' to you?"

"It could be a ploy," Bran insisted, though uncertainty wavered in his voice. "We can't rule anything out."

Edrin shot them all a warning look, voice low. "Enough. We'll stay alert. But right now, she's our best chance." Practical as ever, he added, "If she wanted to harm us, there were easier ways than offering a solution we desperately need."

Theryn smirked lightly. "Could be a long play," he muttered, "After all, I know all about those." He let it drop with a shrug when Lysera gave him a quelling stare.

Within minutes, Alira guided them down a side stairwell off a quiet courtyard, far from Threadmere's bustling forums. Underneath the city, the air turned cool and heavy with the scent of ancient stone and lichen. They descended a winding corridor lit only by Alira's small hand-held glyph-lantern, its blue light casting dancing shadows on walls etched with faded runes. Bran sniffed and caught a whiff of old dust and something vaguely sweet—perhaps residue from long-dormant spells. It tickled his nose, and he sneezed softly, earning a sympathetic half-smile from Alira over her shoulder.

At last, they arrived at a sealed door of intricately carved wood and brass. Alira placed her palm on a central glyph marking and whispered a few words. The lock released with a rusty click. "Mind your step," she cautioned as they entered the hub chamber.

The room beyond was circular and cavernous, an abandoned nexus deep beneath Threadmere. The floor was inlaid with concentric rings of copper and silver glyph-work, dull with age. In the center lay a stone dais, also graven with symbols that curved like whorls of thread. Lysera's breath caught; she could feel the hum of old magic here, a faint thrum in her chest like the echo of a heartbeat.

Edrin and Theryn fanned out instinctively, checking the shadows for any lurking danger, but there was only silence and the sense of heavy, ancient power. Bran walked gingerly along the edge of the central circle, eyes wide in wonder. "So this is a Keylie Path hub… Incredible. I never thought I'd see one."

Alira moved with practiced efficiency. She shrugged off her cloak and draped it over a broken pillar. Underneath, she wore simple travel-stained robes marked with a few embroidered glyphs at the cuffs. Kneeling at the dais, she began to inspect the grooves. "It's been a while since anyone used this," she murmured. She pulled out a small vial from her satchel and poured a shimmering teal liquid into a shallow indentation in the stone. The glyph patterns came alive, light rippling out in pale lines from the center as the conduit eagerly drank the offering. "Leyline ichor," she explained in response to the unasked question, swirling the remaining contents of the vial. "A catalyst to awaken the glyphs."

Theryn hovered near Edrin, hand on his sword hilt as he eyed the brightening glyphs warily. "This place feels… on edge," he muttered. "Like it's waking up and might not be too happy about it."

"Oh, that's normal," Alira assured gently. She stood and faced them, the now-glowing pattern casting moving light across all their faces. "I'll open the path and guide you through. I need all of you to stand within that inner ring." She pointed to a circle etched on the floor around the dais. "When the way opens, you may feel… strange. It isn't a simple teleport. Your minds will travel the leyline route even as your bodies are carried along. You must keep hold of yourselves and *stay together*."

Edrin nodded and ushered the others into the indicated ring. "What about you?" he asked. "You speak as if you aren't coming with us."

Alira offered an apologetic half-smile. "I'm not. The Keylie Paths need a guide at the origin to maintain the connection. I will channel the energy from here and essentially *push* you along the leyline to the Red Kiln hub." She saw Edrin's concerned look and added, "Don't worry, the hub on the other side will slow you safely… if all goes well."

Lysera laid a hand on Alira's arm briefly. "You're taking on a great risk for us. Thank you." The young adept was trembling ever so slightly—whether from fear or the effort of gathering her magic, it was hard to tell.

Alira managed a soft chuckle. "Thank me by stopping Malrik. I have my reasons, believe me." She released a slow breath, grounding herself. "Other threats are brewing. I need to stay—

more mirror-tainted artifacts might surface, and if someone *is* influencing Malrik, I aim to uncover it from this end. That's why I can't go with you. But I won't abandon you either. I'll be right here, holding the leyline open until you're through."

Bran looked at Alira earnestly. "We suspected, um... Well, some of us wondered if you might be..." He trailed off, suddenly feeling awkward now that she had shown such earnest resolve.

Alira tilted her head. "Wondered if I might be what?"

Theryn cleared his throat. "Evil," he said bluntly, shooting Bran a look. "We wondered if you might be secretly evil. No offense."

To their surprise, Alira laughed, a short, genuine laugh echoing off the stone chamber. "None taken. Frankly, if our roles were reversed, I'd be wary too. But I assure you, I want Malrik stopped. He's hurt people I care about... and he's delving into magics that could unravel a lot more than one town." Her expression darkened at that last thought, but she quickly composed herself.

Edrin gave a thin, apologetic smile. "These days, we've learned to question coincidences. But, for what it is worth, I believe you, Alira. And we're grateful."

With that settled, Alira motioned for them to join hands around the dais. The four companions formed a small circle, and Alira stepped just outside the ring, rolling up her sleeves. She began to chant under her breath, tracing complex sigils in the air with her fingers. In response, the glyphs on the dais flared brighter, lines of light rising from the floor to spiral around them like luminous threads. A deep vibration filled the room.

Underneath her clothing, along her shoulders, something moved, twitching under the simple cloth. Theryn hissed, "Fairykin!"

Alira smiled, "Yes, and? Who better to know the fairy pathways?" She began her ritual.

Lysera's eyes fluttered closed as the Loom's energy brushed her senses; she felt a chorus of distant emotions, like echoes of hundreds of past travelers lingering in the path. Edrin tightened his grip on Theryn's and Lysera's hands, grounding himself as the floor seemed to fall away into glowing infinity beyond the circle. Bran couldn't resist peeking—a mix of terror and wonder on his face—as he saw the walls of the chamber fade, replaced by a void pricked with innumerable points of blue-white light. It was as if they stood in the center of the night sky itself, on a platform woven of starlight.

Alira's voice rose, clear and commanding now, reverberating with power: "Hold fast to each other. Focus on Red Kiln—on why you must get there." She stepped forward and pressed her palms flat against the outside of the copper ring. Her eyes had turned bright with raw magic, silver irises luminescent. "Feel your intent. That emotion will guide the leyline to your destination."

As instructed, each of them summoned the driving emotion that brought them here: Edrin's determination to prevent the catastrophe he'd seen before in war; Lysera's compassion for the innocents in harm's way and fear of Loom threads snapping; Bran's curiosity and deep need to see wrongs righted; Theryn's loyalty to his friends, masked as usual by a joking bravado, and his urge to face the danger head-on so others wouldn't have to. The leyline current responded. Threads of light coiled around their feet and waists, ready to launch them across the world.

The chamber around them began to blur. Alira's form wavered at the edge of their vision, her lips moving in a final entreaty they could barely hear. "May the gods weave with you," she whispered, voice strained with the effort of holding open the roaring magical torrent. With a sharp whoosh, the Keylie Path yanked the group forward.

For an instant, each of them felt weightless—consciousness stretching like a filament caught in a gale. Edrin thought he heard a wolf's distant howl and saw a flash of ghostly blue eyes in the dark; Bran smelled wet earth and lavender as images of an endless library flickered by; Theryn swore he heard laughter—his sister's?—echoing from the stars; Lysera felt a tear on her cheek as she glimpsed a towering loom threaded with countless constellations. Then, just as quickly, the sensations were gone, replaced by a single, jarring pull.

Their bodies vanished from the Threadmere hub in a twist of light. Alira staggered back, suddenly alone in the silent chamber. The glyph-light dimmed to a faint glimmer underfoot, the ritual complete on her end. Chest heaving, she sank against the stone dais, praying under her breath that they would reach their target safely.

Far along the leyline, hurtling unseen through Aurelia's mystical veins, Edrin, Lysera, Theryn, and Bran traveled as fast as thought. Their fate now lay in the balance of the Keylie Path and the mercy of the Loom. And as Alira pushed herself upright to begin the hard task of anchoring their route from afar, she could only trust that this desperate gamble would deliver the heroes in time to save Red Kiln from the encroaching darkness.

- CHAPTER NINE -

Red Kiln

Chapter Nine

Red Kiln – Afternoon Lioraday, 23 Solaryn's Crown, TT 1215

The Keylie Path spat them out like reluctant pits found in a fruit. Their bodies materialized upon the scorched hilltop in a frenzy of light, collapsing as quickly as it had flared. For one heartbeat, they existed in both places—minds still trailing through the luminous corridors between worlds even as their boots struck solid ground. Then reality asserted itself with brutal finality, their consciousness settling back into flesh and bone. The Red Kiln spread before them, its brick circle etched against the empty sky like an accusation, the earth around it transformed to glass, capturing and bending the harsh sunlight into fractured rainbows.

Edrin staggered forward, his vision swimming with afterimages of stars and darkness. His legs felt simultaneously leaden and weightless, muscles remembering a journey they hadn't physically taken. Behind him, Theryn cursed colorfully as he dropped to one knee, fingers splayed against the ground to steady himself. Lysera remained standing but swayed like a reed in a gentle current, her eyes unfocused as she processed the transition. Only Bran seemed relatively composed, his scales flexing and contracting even though his movements were slower than usual.

"Everyone whole?" Edrin asked, voice rougher than intended. He counted limbs and companions with the automatic vigilance of someone accustomed to emergencies.

Theryn rose, brushing glass dust from his palms. "Mostly," he replied, rotating his shoulders with a grimace. "Feels like my mind arrived slightly before the rest of me. Uncomfortable arrangement. I find myself not wishing to repeat that journey."

The earth beneath their feet had been transformed by extreme heat, normal soil vitrified into glass-like plates that crunched underfoot. Heat still radiated upward, making the horizon waver. Edrin advanced toward the Red Kiln's central ring, hammer already in hand, eyes scanning for movement, for preparations, for any sign of Malrik or his ritual. The brick structure pulsed dully in the sunlight, its reddish surface breathing with residual heat, but nothing moved within its circular embrace.

"Spread out," Edrin commanded, his voice carrying the edge of a man prepared for ambush. "Theryn, check the perimeter. Bran, look for fresh glyphs. Lysera…"

"I'll read what's left," she finished for him, her fingers already tracing invisible patterns in the air before her.

They moved with practiced efficiency despite their lingering disorientation, each taking a section of the site. Theryn prowled like a predator testing unknown territory, his steps liquid and silent despite the brittle surface. Bran crouched at intervals, scales shifting hue as he studied patterns burned into the glass. Lysera stood with arms outstretched, palms open as if feeling for a current invisible to ordinary senses.

Minutes stretched into a quarter-hour, and still they found nothing. No ritual implements. No sacrificial altar. No glowing sigils awaited activation. Just emptiness and the lingering sense that powerful forces had recently departed.

"There's nothing here," Edrin finally admitted, the words tasting of defeat. "No Malrik. No ritual. Nothing but burnt earth and dead air." His voice cracked slightly on the last words, betraying the depth of his frustration.

Theryn returned from his circuit, disappointment evident in his uncharacteristically straight posture. "Not even bootprints beyond our own," he confirmed. "If anyone was here, they're long gone."

Edrin's control, maintained through days of forced marches, sleepless nights, and the desperate dash through the Keylie Path, finally fractured. With a wordless growl, he slammed his fist against a nearby rock, the impact sending shards of obsidian skittering across the vitrified ground. Pain shot through his knuckles, but the physical hurt was almost welcome. His shoulders bunched beneath his tunic, jaw clenched tight enough that tendons stood out along his neck. Behind his eyes burned the memory of Emberfield in ruins, of bodies covered with makeshift shrouds, of flames that consumed everything that mattered.

"We were so sure," he said, voice low and dangerous. "So certain this was where it would happen."

Lysera approached cautiously, her own disappointment evident in the droop of her shoulders, yet her eyes remained clear. "The energy here is…strange," she offered. "Like an echo of what should be present. I can sense where power gathered, but it's as if someone scooped it away, leaving only the impression behind."

While Edrin wrestled with his anger and Lysera attempted to read the invisible residue of magic, Bran had wandered toward the kiln's eastern arc. There, partially obscured by a drift of glassy sand, he discovered a half-burnt glyph etched into the ground. Unlike the natural patterns formed by heat and pressure, this marking showed deliberate design reflecting intelligence rather than accident. He knelt beside it, brush in hand to clear away debris, his scales darkening with concentration.

"Edrin," he called, voice carrying that distinctive tone scholars reserve for significant discoveries. "You need to see this."

The others converged on his position, forming a loose semi-circle around the uncovered glyph. Burned into the ground was a symbol unlike the standard fire-glyphs associated with Pyrrhos or the protection wards they'd encountered at other sites. This marking resembled a fractured eye, its pupil replaced by a miniature labyrinth that seemed to shift when viewed from different angles.

"Mirrathar work," Bran pronounced, his tone both respectful and wary. "Mirror-magic, but used as a placeholder rather than as active spellwork."

Edrin crouched beside him, anger temporarily subsumed by curiosity. "What does it mean?"

"It means," Bran replied carefully, extracting a cloth pouch from his pack, "that this entire site was prepared as a decoy." With delicate precision, he collected a mirror fragment, wrapping it in the cloth before securing it in an inner pocket. "The pattern is designed to hold attention but not power—to appear as the genuine article when viewed through most magical senses."

Theryn whistled softly, the sound cutting through the heavy air. "Someone went to a lot of trouble to make us think this was important." His eyes narrowed as he scanned the horizon, as if expecting to spot the architect of their misdirection lurking just beyond sight.

Lysera joined them, her expression grave as understanding dawned. "The counting changed," she said softly. "Just before we left Threadmere. I thought it was acceleration of the ritual, but now I see, it was a different rhythm altogether. A different working."

The four companions stood in a tight circle, the realization settling over them like the oppressive heat radiating from the ground. They had been drawn away, sent racing across the continent to investigate a threat that was real but misplaced. A shadow box while the true performance unfolded elsewhere.

"Someone wanted us away from Threadmere," Bran concluded, rising to his feet. "Far away from where we were actually needed."

Edrin's face hardened into grim determination as he stared back the way they had come, though the Keylie Path left no visible trail to follow. "The mirror witch," he said, the title falling like a stone between them. "She played us like festival puppets."

Lysera nodded, her intuition aligning with his logic. "The council meeting was too convenient, her opposition too calculated. She needed us gone."

"And Alira?" Theryn asked, the question hanging in the air.

"Genuine, I think," Bran answered after a moment's consideration. "But perhaps manipulated as well. The mirror-shard has Mirrathar resonance, but corrupted, twisted away from its true purpose."

Edrin looked toward the horizon, where heat haze still blurred the earth and sky. His frustration had crystallized into something harder and more focused—a determination burning cold rather than hot.

"We need to get back to Threadmere," he said, the words simple but heavy with implication. "Whatever they wanted us away from is happening there, not here."

The others nodded in agreement, already gathering what little they had unpacked. The Red Kiln stood silent behind them, its empty ring a circle of broken promises and misdirection. Ahead lay uncertainty and the daunting task of retracing their steps across a continent. But now they moved with clarity, knowing that they had been deceived, and with the resolve after such enlightenment.

#

Leyline travel exacted its toll with merciless precision, drawing payment in trembling muscles and frayed nerves rather than coin or favor. They made camp as twilight gathered around them, a safe distance from the Red Kiln where residual energies still pulsed beneath the glass-fused earth. Each companion carried the journey differently: Theryn's usual fluid grace had acquired a subtle stutter, like a clockwork mechanism with one gear slightly bent; Bran's scales had dulled to the flat sheen of river stones in drought; Lysera's movements remained graceful but measured, as if she calculated the energy cost of each gesture. But it was Edrin who bore the heaviest burden, the Keylie Path having carved temporary channels through his being, collapsed inward, leaving him with pain that flowed like molten metal behind his eyes.

He had insisted on helping to establish their meager camp—stubborn pride compelling him to stack firewood and arrange bedrolls despite the tremors that occasionally seized his hands. Only when the essential tasks were completed did he finally surrender, dropping heavily onto a flat stone and pressing his palms against his temples. The headache pulsed in time with his heartbeat. His vision blurred at the edges, colors bleeding into one another like watercolors left in rain. Muscles that had carried him through battles and forging work now cramped without warning, phantom fingers squeezing his calves and forearms with unpredictable malice.

Lysera watched him from across the clearing where she had been arranging herbs for a simple meal. His discomfort was written in the rigid line of his shoulders, in the tight compression of his lips, in the way his eyes squeezed shut against sensations invisible to others. Her own body ached from the journey—joints stiff and head mildly throbbing—but her symptoms were mere whispers compared to the screams of Edrin's suffering. Without conscious decision, she abandoned her task and moved toward him.

"Edrin," she said softly as she approached, not wanting to startle him when his defenses were already compromised. "I have heard that the Keylie Paths are always hardest on those with the strongest physical connection to the earth. Your body is fighting to reclaim its natural place in the world."

He looked up, the movement careful and controlled, as if his head might shatter with any sudden motion. The pain in his eyes triggered an answering ache in her chest—not sympathy alone, but something deeper and more personal.

"It'll pass," he managed, the words emerging through clenched teeth. Even now, he attempted to present strength where vulnerability would be more honest.

Lysera ignored his stoic dismissal, positioning herself before him with gentle purpose. "May I?" she asked, hands extended but not yet touching him, respecting the boundary of his personal space despite her desire to help.

Edrin hesitated—permission for touch, for help, for weakness all tangled in the simple question. Then, as another spasm wracked his shoulders, he nodded once.

She guided him to lean against a smooth boulder, its surface still warm from the day's heat. Then she knelt before him, the position both practical and strangely intimate, her knees pressing into the sandy soil. Her hands reached for his, enfolding his calloused palms between her slender fingers. His skin felt fever-hot, the muscles beneath bunched into knots she could feel even through this simple contact.

"Breathe with me," she instructed, her voice taking on a subtle resonance that echoed despite the open air around them. She demonstrated, drawing air deep into her lungs from both her mouth and nose, and releasing it in a slow, controlled stream only through the mouth. "Like this. In through the nose and mouth, filling your belly first, then your chest. Out through the mouth, emptying completely."

Edrin tried to match her rhythm, but his first attempts were shallow and irregular, his body resisting the surrender that healing required. Lysera maintained her steady pace, neither hurrying nor showing impatience. Her thumbs traced small circles against the base of his palms, finding pressure points that connected to pathways throughout his body. With each circuit, she loosened another knot of resistance, another barrier to relief.

"The Keylie Paths aren't natural for human bodies," she explained, her voice low and soothing, giving his mind something to focus on besides pain. "We weren't meant to travel that way. Our spirits can flow through the Loom's threads, but our bodies resist being pulled apart and reassembled. What you're feeling now is your essence remembering its proper shape."

As she spoke, her fingers worked upward from his hands to his wrists, then along his forearms, tracing the paths of blood and energy beneath his skin. The air between them warmed with each touch, not with embarrassment but with the honest heat of connection.

Gradually, Edrin's breathing deepened to match her own. The rigid line of his shoulders softened by increments, muscles remembering how to release rather than brace. His vision cleared, the world's edges sharpening from their previous blur. The headache remained, but dulled from jagged agony to manageable discomfort. He was astonished. All of that power in one woman. She was truly magnificent, he thought.

A strand of hair had fallen across his forehead, dampened by sweat and clinging to his skin. Lysera reached up without thinking, brushing it back with a touch so light it might have been mistaken for the evening breeze. Her fingers lingered at his temple, feeling the strong pulse of blood beneath the skin, the vital current that connected him to the world's greater rhythms. For a moment, she forgot her purpose, caught in the simple wonder of this contact—fingertips against warm skin, life recognizing life.

Edrin's eyes opened fully, meeting hers with unexpected clarity. In that gaze lay not just gratitude but recognition—the wordless acknowledgment of something that had been growing between them through shared dangers and quiet moments. Neither spoke, neither needed to. The silence between them held more meaning than fumbling words could capture.

Time seemed to pause around them. The world narrowed to this small circle of connection. Lysera became aware of details with preternatural sharpness: the faint scar that bisected his left eyebrow, the flecks of amber in his otherwise dark eyes, the way his pulse visibly beat at the hollow of his throat. The rough stubble of the beard that he rarely shaved, yet hardly grew to more than just a soot-like covering of his face. She felt caught in his gravity, drawn toward him by forces as natural and inevitable as the tide pulled by the moon.

Edrin's hand lifted slowly, muscles now obeying his will rather than rebelling against it. His calloused palm found her cheek, the rough texture against her smooth skin, the contrast emphasizing the distinct paths they had walked to arrive at this shared moment. His touch carried both strength and gentleness—the duality that had first drawn her attention when he was still a stranger rather than whatever he had become to her now.

"Thank you," he said, the words simple but weighted with meaning.

Lysera leaned almost imperceptibly into his touch, allowing herself this small acknowledgment of desire. "We carry each other," she replied, her voice barely above a whisper. "That's how we survive. If we stay together, we can stop this craziness and get back to a normal life."

They remained thus connected for several heartbeats more, neither willing to be first to break the fragile bridge between them. Then, from across the camp, Bran cleared his throat—not from disapproval but from practical recognition that food needed preparation and plans required discussion. The spell gently dissolved, though the warmth it had generated remained, a banked fire that continued to provide comfort even as they slowly separated and turned toward more mundane concerns.

Edrin rose with new steadiness; the worst of the leyline fatigue pushed back by Lysera's care. The headache had receded to a dull pressure rather than sharp pain, and his vision was now clear enough to navigate the gathering darkness. Yet even as he moved to help with the evening meal, his hand briefly found her shoulder.

Lysera followed, knowing that something essential had shifted between them—a recognition too profound to dismiss yet still too new to name. The intimacy they had shared would remain, a foundation upon which whatever came next would build itself, stone by careful stone.

#

The evening descended in gradual layers of indigo and shadow, stars kindling one by one, threaded through the fabric of night. Bran crouched beside the small fire he had coaxed to life. His scales reflected the dancing flames in muted copper patterns while his clawed hands crumbled dried herbs into the pot of water, just reaching a simmer. Nearby, Theryn sliced strips of preserved meat with methodical efficiency, the blade of his knife catching firelight with each stroke. Both maintained the appearance of complete absorption in their tasks, yet their peripheral awareness remained fixed on the tableau unfolding across the camp—Lysera kneeling before Edrin, their hands joined, their gazes locked in silent communion.

Theryn nudged Bran with his elbow, the gesture so subtle it barely disturbed the lizardfolk's careful addition of salt to the pot. With a tilt of his chin, he directed Bran's attention toward the pair across the clearing. "Would you look at that," he whispered. "Took a misdirection, a Keylie Path, and near-fatal headaches to get them that close."

Bran's scales rippled with amusement, though he maintained his focus on the broth, adjusting its consistency with practiced motions. "Some humans require extraordinary circumstances to acknowledge ordinary feelings," he observed, his tone carrying the patient wisdom of one who had observed many such dances over his longer lifespan.

When Edrin's hand finally rose to touch Lysera's cheek, Theryn suppressed a triumphant grin, instead directing it into the steady rhythm of his knife against the wooden cutting board. "Field-tested romance," he whispered to Bran, the teasing lightening his features after days of tension and disappointment. "Nothing like a wild goose chase to bring hearts together."

Bran stretched his neck, scales catching the firelight as he worked out a stiffness that had settled in since their journey through the Keylie Path. A small wince escaped him as he rotated his shoulders. "If they'd just admit what's obvious to everyone else, we could all get some rest," he replied, but the fondness in his tone belied any real irritation. His violet, vertical pupils showed genuine warmth as he glanced toward their companions. "Though I suppose some patterns must unfold at their own pace."

Theryn snorted softly. "Their pace would exhaust a glacier's patience." He emptied the sliced meat into the pot Bran tended, the strips curling as they met the hot broth. "Want to wager on who speaks first? My coin's on Edrin—he'll blurt it out during some crisis, no poetry whatsoever."

"I take no bets where human hearts are concerned," Bran answered primly, though a smile tugged at the corners of his scaled mouth. "The Loom weaves those threads too unpredictably."

Their quiet conversation paused as Bran cleared his throat. The two companions separated with visible reluctance, like plants that had been growing toward the same light suddenly finding themselves in different soils. As they approached the fire, Theryn and Bran exchanged a final knowing glance before assuming expressions of innocent concentration on the meal.

"Something smells almost edible," Edrin remarked as he settled on a flat stone near the fire, his voice stronger now, though fatigue still shadowed his features. "A miracle, considering our supplies."

Bran ladled the simple broth into their travel cups, steam rising in fragrant spirals. "The miracle is that Theryn hasn't complained about the lack of seasoning yet," he observed, passing the first portion to Lysera with a slight bow of his head.

"The night's still young," Theryn quipped, accepting his share with a flourish. "I'm saving my critique for after I've consumed every last drop."

As they ate, the teasing continued in subtle currents beneath their conversation—a raised eyebrow when Edrin automatically shifted to make room for Lysera beside him, a poorly hidden smile when their hands accidentally brushed while reaching for the same piece of flatbread. Neither Bran nor Theryn commented directly, but their eyes carried messages clear enough to anyone paying attention. Lysera caught Theryn mid-smirk once and shot him a withering look. The gentle mockery persisted, a welcome distraction from the greater concerns that surrounded them.

Their meal, meager though it was, restored a measure of strength to bodies still recovering from the Keylie Path's toll. As they passed the waterskin between them, their conversation inevitably turned toward what awaited them in Threadmere and how they might reach it.

"The Keylie Path can't be used again so soon," Bran explained, tracing patterns in the dirt with a claw as he spoke. "Even if we knew how to activate it from this end without Alira's help, the strands need time to recover, to reweave themselves after being stretched."

Edrin nodded, his expression thoughtful rather than defeated. "Then we travel conventionally. If we can secure mounts from the nearest settlement, we could reach Threadmere in—what? Two or three weeks?"

"A little under two weeks on foot," Theryn confirmed. He had set aside his empty cup and now reclined against his pack, apparently relaxed, though his eyes remained alert. "Or less, assuming we find horses willing to carry us after they catch our scent." He wrinkled his nose dramatically. "Leyline travel leaves a distinct aroma that animals find…concerning."

Lysera smoothed her skirts, a habitual gesture that brought order to at least one small corner of their uncertain world. "I'm more concerned about what we'll find when we return," she said, her voice soft but clear. "Whatever the mirror witch aimed to accomplish by removing us from Threadmere, she's had time now to set it in motion."

A somber silence followed her words. The fire popped and hissed, sending a spray of sparks toward the star-filled sky. Each spark lived its brief, brilliant life before fading back to darkness—a reminder of how quickly light could be extinguished, how precious each moment of brightness.

"The Council was divided over our and Vaelis's proposals," Bran noted, his scales dimming slightly as he recalled the charged atmosphere of the chamber. "But with us gone, with no counter-voice remaining…" He left the thought unfinished, though its conclusion hung in the air between them.

"She'll still need a vessel," Theryn pointed out. "Someone willing to accept concentrated divine power. That's not a simple matter of politics or persuasion."

Edrin's expression darkened. "Unless she already has someone. Unless this was never about convincing the Council, but about keeping us from seeing who she'd chosen. Or, maybe it is just her. Maybe Vaelis is the mirror mage, and she is vying for a mantle of some god."

The implication settled over them like a physical weight. If their opponent had already selected her vessel and begun channeling divine power into a single mortal frame, then their return journey became a race against distance and transformation itself.

"We need more information before we can be sure," Lysera cautioned, ever the voice of measured thought. "But whatever awaits us, we face it together." Her hand found Edrin's arm, a brief touch that carried strength rather than comfort. He acknowledged it with a nod, his jaw set in lines of determination.

As preparations for sleep began, Edrin rose and walked to the camp's edge. He stood silhouetted against the night sky, his posture straight despite the day's trials, his gaze fixed on the distant horizon where Threadmere lay beyond sight but not beyond thought. The stars above him seemed to pulse with ancient rhythm, a heartbeat too vast for mortal comprehension yet somehow tied to their small struggles.

Theryn approached, moving with his characteristic silence. "Planning the route?" he asked, though he knew the answer extended beyond simple geography.

"Planning our return," Edrin replied, his voice steady with renewed purpose. "Whatever they wanted us away from, we're going to find it. Whatever game that manipulator is playing, we're going to end it." His hand fell to the hammer at his belt, fingers curling around the worn handle. "She thought removing us would clear her path. She'll learn how wrong she was."

Theryn nodded, his usual sardonic expression replaced by something rarer and more genuine—respect untainted by irony. "I'd almost feel sorry for her," he said, "if she weren't trying to unmake the world."

The two men stood side by side in companionable silence, their shadows stretching long behind them in the firelight. Behind them, Bran and Lysera prepared bedrolls and set the watch rotation, their movements synchronized by days of shared travel. Together, the four companions formed a constellation of their own.

As night deepened around them, Edrin remained at his post, sentinel against the darkness. His thoughts ran ahead of him to Threadmere, to whatever awaited them there, to the confrontation that now seemed inevitable. The path forward was neither clear nor easy, but it was necessary, and in that necessity, he found resolve enough to face whatever lay ahead.

- CHAPTER TEN -

Threadmere

(The Mirror Witch's Chamber)

Chapter Ten

Threadmere – Evening (Pyrrday, 24 Solaryn's Crown, TT 1215)

Beneath Threadmere's winding streets and oblivious crowds, the practitioner's secret chamber crouched in darkness, its stone walls absorbing all sound save what she chose to admit. Candle flames wavered in currents of air too subtle for ordinary perception, casting unnatural shadows across surfaces designed to catch and bend their light. At the chamber's heart sat the mirror mage herself, naked as usual for her rituals. She admired herself in all of the mirrors, sitting cross-legged before her hanging pendant, its fractal surface drinking in the sparse illumination, transforming it into geometries that eyes were never meant to trace.

Her chamber had been carved beneath the foundations of a temple long since repurposed, forgotten corridors sealed by stonework that remembered its original purpose even when minds did not. Seven candles formed a perfect circle around her seated form, their flames burning in colors oscillating between silver, blue, and the faintest hint of void-black at their centers. Ancient sigils, scratched into the floor with tools fashioned from materials that had never known the forge, pulsed with subtle luminescence whenever the candlelight touched them at precise angles.

The mirror mage's light skin seemed to absorb rather than reflect the chamber's meager light. Her hair moved with liquid grace around her shoulders, blue and silver tendrils occasionally rising against gravity's insistence when her power surged. Her hands hovered over the pendant, fingers splayed like the legs of some exotic arachnid, skin catching the fractal light in patterns with impossible geometry. She was beautiful in the way that abyssal creatures are beautiful—an elegance born of perfect adaptation to environments hostile to ordinary life.

From somewhere far above, the sounds of Threadmere's evening life filtered down through stone and earth. A merchant's cart rattled over cobblestones, its wheels striking a loose stone, sending a high, clear note echoing through hidden cracks. A town crier called the hour, his voice reduced to a rhythmic murmur of indistinct syllables that nonetheless carried the cadence of official pronouncement. Somewhere, a child laughed and was answered by a mother's call—the sounds of ordinary joy traveling through stone that should have been impossible. A flock of pigeons took flight from a nearby bell tower, their wingbeats creating a subtle percussion against the chamber's quiet.

The illusionist allowed these sounds entry, not from sentiment but from pragmatism. The boundary between her hidden world and the ordinary one above required careful maintenance. Too much separation could make her forget the patterns of common life needed to maintain her disguise; too little, and she exposed herself to detection. So she listened to the city's breath while she worked, incorporating its rhythms into her ritual like a composer weaving counterpoint through a complex fugue.

Her eyes closed as her fingers found the pendant's edge, tracing its perimeter with the precision of a mathematician. The mirror-focus responded to her touch, its surface rippling like disturbed mercury before settling into new configurations.

Geometries folded inward, revealing dimensions that existed perpendicular to ordinary space, angles that could not exist in reality yet somehow persisted when given form through her will.

"Show me the dreamer," she whispered, and the pendant obliged.

Within its fractal depths, Malrik's sleeping form appeared—as a projection existing simultaneously at multiple depths within the mirror's impossible space. He slept fitfully in some mountain cave, body occasionally twitching as dreams already troubled his rest. Perfect, she thought. The agitated mind accepts new patterns more readily than the peaceful one.

The mage's fingers danced across the mirror's surface, leaving trails of intention rather than physical marks. The pendant absorbed these patterns, translating will into vision with the efficiency of a language evolved specifically for this purpose. The first image took shape within the depths: Malrik seated upon a throne fashioned from blackened bone and glowing ember. The bone components shifted subtly, suggesting the remains of those who had opposed him, while the ember elements pulsed with a heat that scorched without consuming—eternal punishment without the release of destruction.

In the vision, crown-wearing dignitaries knelt before him, their faces bearing expressions of fear mingled with adoration. Malrik's dream-self smiled, a gesture of perfect satisfaction. The sorceress refined the image, adding details that would speak to his specific hungers—the trembling hands of a deposed king offering up his scepter, the downcast eyes of once-proud nobles now reduced to supplication. The throne room itself she crafted with architectural precision, its dimensions suggesting both boundless power and intimate control.

"This could be yours," she whispered to the mirror, her words traveling paths that physical sound could never follow. "This will be yours."

The pendant absorbed the first vision, storing it in facets that multiplied its impact through repetition across slightly varied iterations. The ritualist moved to the second tableau without pause, her fingers now drawing more complex patterns as sweat began to bead at her temples.

Within the mirror's depths, Malrik stood in a circular chamber reminiscent of the Red Kiln, though transformed by his ascension. Prisoners knelt before him, their spirits not yet broken but poised at the exquisite moment before surrender. In his hands, Malrik held metal rods tipped with living flame that responded to his will, extending or withdrawing their heat according to his whim. The prisoners' faces reflected fear and the dawning recognition that resistance was futile against a power transcending mortal limitation.

"They will yield to your flame," the mirror witch murmured, her voice taking on harmonics that caused the candles to flicker in sympathy. "All who opposed you, all who doubted, all who failed to recognize your destiny. They will know the truth through pain."

The mirror drank in this vision as it had the first, its facets multiplying the scene into countless variations that would assail Malrik's dreaming mind from every conceivable angle. The watcher felt the first genuine strain of the working, a pressure behind her eyes as she shaped the third and final vision. Her reflection fragmented across the pendant's surface, splitting into shards with each containing a slightly different version of herself—some more human, others less so, all watching her work.

The third vision bloomed within the mirror: Malrik surrounded by bodies yielding to his pleasure. Some came willingly, drawn by the irresistible gravity of his power; others resisted until resistance itself became another form of submission. Naked and kneeling before his powerful naked form, his manhood swollen and erect, actions of pure torture becoming pleasure, not pain. The veiled adept crafted this vision with particular care, weaving threads of domination and worship together until they became indistinguishable. The sensual tableau shifted constantly, male and female figures merging and separating, speaking to desires Malrik himself might not consciously acknowledge but would recognize in the vulnerable state between sleeping and waking.

"Power and pleasure," she whispered, her voice now strained as the working neared completion. "Dominion and desire. They are yours to take, yours to command."

Sweat now ran freely down her temples as the pendant pulsed with accumulated intent. Her reflection had splintered completely, each fragment showing a different aspect of her being—some beautiful beyond mortal comprehension, others terrible in their alien symmetry. The candles' flames stretched toward the ceiling, elongating into impossible columns of light that cast no illumination yet somehow defined the chamber's dimensions more precisely than ordinary brightness could.

The mirror-worker gathered her remaining strength for the final sending, pressing both palms against the pendant's surface. The mirror-focus responded by becoming momentarily permeable, allowing her fingers to sink partially into its substance as if dipping into a pool of liquid potential.

"Rule with me," she whispered, her voice carrying the full weight of her will, "and all Aurelia will kneel before us."

The pendant flashed once. An intense burst of light momentarily erased all shadow from the chamber, rendering everything in stark black and white. Then darkness returned as the mirror dimmed, its surface cooling from molten possibility back to dormant potential. The visions had been sent, carried across miles to Malrik's dreaming mind, where they would take root like seeds planted in fertile soil.

The practitioner slumped forward, momentarily exhausted by the effort. But even in fatigue, her lips curved into a smile of satisfied anticipation of a trap well-baited. Malrik would believe the visions came from his own ambition, his own destiny. He would never suspect that the very power he thought would elevate him was merely the chain by which she would bind him to her true purpose.

Above her, Threadmere continued its evening routines, citizens going about their lives unaware of the currents shifting beneath their feet. A bell tolled the late hour, its sound reaching even the mirror mage's hidden chamber like the heartbeat of a world soon to be remade.

#

The Council Chamber of Threadmere wore its centuries of authority like a well-tailored cloak, its high marble ceiling collecting whispers and proclamations alike into a patina of power breathing with the building itself. Sunlight poured through tall stained-glass windows depicting the city's founding moments, fracturing into jewel-toned patterns across the polished floor where petitioners had stood for generations seeking judgment, mercy, or simply to be heard. Heraldic banners hung from iron brackets between the windows, their dragons and thread-motifs rendered in gold and silver thread, catching light when crossdrafts from opening doors made them flutter like living things. The chamber smelled of beeswax candles, leather-bound books, and the subtle perfumes worn by councilors who believed power should have a pleasing scent.

The assembled councilors once again occupied tiered seats of carved oak, forming a half-circle around the speaking floor. Their robes—variations on the city's colors of deep blue and silver—rustled with subtle importance as they shifted position, consulted notes, or leaned to whisper commentary to their neighbors. Behind them, apprentice scribes perched on smaller seats, quills poised over parchment as they recorded the day's proceedings with expressions of studied neutrality. In the gallery above, minor nobles and guild representatives jostled for better viewing positions, their occasional murmurs creating an undercurrent beneath the more formal proceedings below. A group of visiting merchants from the eastern provinces stood out in their bright yellows and oranges, their expressions betraying unfamiliarity with Threadmere's more somber protocols.

Before this assembled authority stood Edrin Ashwood, his shoulders squared and jaw set in the determined posture of a man who had faced greater dangers than disapproval. Weeks of hard travel had left their mark on him—his clothes, though clean, showed the creases of hasty washing, and his face showed insufficient rest, while a new scar across his left knuckles testified to encounters beyond the council's protected walls. Beside him stood Lysera, her normally serene features tightened with the effort of maintaining composure despite fatigue. Her sight-veils—gossamer strips of fabric embroidered with seeing-glyphs—hung from her wrists, occasionally trembling when particularly disturbing visions passed through her awareness.

"Councilors of Threadmere," Bran began, his voice carrying the burr of Emberfield's accent more prominently than usual, a sign of his emotional state. "We return with evidence of deliberate misdirection. The threat at Red Kiln was fabricated—a decoy designed specifically to remove us from this city." Edrin's hand moved to his belt, where his hammer hung as a grounding presence. Bran continued, "Mirror-fragments found at the site carry the same corruption signature as artifacts previously connected to Mirrathar workings."

Outside, a hawker's cry penetrated the chamber's thick walls—something about fresh fish from the northern lakes, the words rendered indistinct, but their rhythm familiar to all who knew Threadmere's street language. Moments later, the clatter of a dropped crate followed by multilingual cursing provided punctuation to Edrin's points. The city continued its commerce regardless of what transpired in the chambers of power.

"The pattern is consistent," Bran continued for Edrin, his voice rising with frustration at the politely blank expressions before him. "Someone with knowledge of mirror-magic has been manipulating events—creating false threats to divert attention while implementing their true design here, in Threadmere itself."

Edrin gestured toward the councilors with a sweep of his arm. "While you debate, they act. While you consider, they implement."

From the scribe's gallery, the scratch of quills intensified as they captured his words. A young apprentice hastily knocked over an inkwell to record particularly vehement phrasing, earning a sharp glare from his master. The small commotion drew several glances from the council, distracting from Edrin's increasingly urgent tone.

Lysera stepped forward, her movement drawing attention from Edrin's intensity. Her fingers trembled slightly as they brushed across her sight-veils, activating glyphs that glimmered briefly before settling into a steady, subtle glow.

"Honorable Council," she began, her voice softer but no less insistent than Edrin's, "I have tracked disruption patterns in the Loom's weave for seventeen days now. They converge here." She lifted her veils, allowing the glyphs' light to cast delicate patterns across her face. "Not at Red Kiln, not at the border, but here in Threadmere's heart. The threads show manipulation—someone gathering power that should remain distributed, concentrating what was designed to flow freely."

Smoke from oil lamps hung in layers near the ceiling, occasionally stirred by drafts from the high windows. From the street below, the muffled sounds of a passing guard patrol provided counterpoint to Lysera's words—boots striking cobblestones in perfect rhythm, armor plates clicking with disciplined movement. A chamberlain entered through a side door, with a pitcher of water and fresh glasses for the councilors, his soft footfalls and the gentle clink of glass punctuating Lysera's mystical warnings.

The Council Head—a woman whose silver-streaked hair was bound in elaborate braids signifying her forty years of service—leaned forward slightly, her expression betraying nothing beyond polite attention. To her right, a merchant-councilor with ink-stained fingers tapped the table in a pattern that suggested either impatience or calculation. To her left, the representative from the Archive district adjusted his spectacles, light reflecting from the lenses momentarily obscuring his eyes.

"These are once again serious allegations," the Council Head noted, her tone measured but cool. "Yet, once again, I hear much theory and little concrete evidence."

Murmurs rippled through the chamber—some voices questioning, others dismissive. In the gallery, a scholar in midnight blue robes shook his head and made a notation in a small book bound in copper wire. Nearby, two guild representatives leaned close, exchanging whispers behind raised hands. The visiting merchants looked increasingly confused, their translator struggling to convey the nuanced skepticism in Threadmere's particular brand of political body language.

It was then that Councilor Vaelis rose from her seat near the chamber's center. Her movements carried a fluid grace with the formal stiffness of her blue and silver robes. Unlike other councilors who shifted and adjusted their attire, she wore her official garments as if they were a natural extension of herself rather than an imposed costume. The pendant around her throat remained carefully concealed beneath layers of fabric, though a particularly attentive observer might have noticed how her hand briefly pressed against it before she spoke, as if confirming its presence.

"While I respect the concerns brought by Scholar Silvertail, Master Ashwood, and Seer Vayesh," she said, her voice carrying a subtle echo that seemed to originate slightly behind her words, "we must exercise caution. Without concrete proof of this…conspiracy, we risk creating the very panic these supposed manipulators desire." Her eyes—a gray so perfect it appeared almost manufactured—swept across the chamber, briefly meeting each councilor's gaze. "Fear is a weapon as surely as any blade. Shall we hand it to our enemies by acting precipitously?"

As she spoke, a draft from somewhere in the chamber's depths caused the nearest banner to billow outward. Outside, a distant drum roll sounded from a guard post, signaling a shift change with military precision. The rhythm vibrated through the chamber's stone foundation, felt more than heard by those present.

The Council Head nodded, satisfaction evident in the slight relaxation of her posture. "Councilor Vaelis speaks wisely," she pronounced, effectively closing the matter. "Without tangible evidence, we cannot commit resources to vague suspicions, however sincerely held." Her gaze returned to Edrin and Lysera, not unkind but unyielding. "Once again, you may continue your investigation, of course, but the Council requires more than patterns and theories before taking action."

Edrin's jaw clenched, a muscle pulsing visibly along its edge. His eyes, shrewd despite his frustration, flicked to Councilor Vaelis, then to the Council Head, and finally to the various officials present. Understanding bloomed in his expression—not just that they had been refused, but that something deeper was at work. They had been outmaneuvered not through argument but through the subtle mechanics of established power, through relationships and influences cultivated long before they had entered the chamber. Now that even Lysera couldn't predict when the next attack or surge might occur, the group was getting desperate.

The heavy doors at the chamber's rear swung open, admitting a gust of air carrying the scents of the city beyond—baking bread from the western quarter, tanning hides from the craftsmen's district, and the ever-present undertone of the river that gave Threadmere both its name and livelihood. Couriers in city livery slipped in, bearing sealed messages that required immediate attention, providing the Council a convenient reason to move to other business.

They were politely but firmly ushered toward the exit. Edrin's hand found Lysera's elbow, steadying her as much as himself. His expression had hardened from frustration to something more dangerous—the look of a man who, having exhausted official channels, was now considering alternatives that lay outside their bounds.

#

The garden alcove clung to Threadmere's upper tier like a bird's nest built into a cliff face, its stone boundaries defining a small sanctuary where time seemed to flow at a different pace than in the city below. Ancient ivy climbed the weathered walls, having conquered the stones generations ago, their gnarled stems thicker than a man's wrist in places and peppered with small white flowers. A flagstone path, worn smooth by centuries of contemplative steps, wound between planters overflowing with herbs whose scents mingled in complex harmony—veilmary's sharpness softened by threadmint, and threadmint's brightness anchored by ashvine. The garden faced westward, capturing the afternoon sun that now painted everything in tones of amber and gold, lending even the most ordinary surfaces a brief nobility before shadows would reclaim them.

Edrin paced the narrow path with barely contained energy, each step landing with more force than necessary. His broad shoulders were bunched tight beneath his travel-worn shirt, hands alternately clenching into fists and splaying open in unconscious mimicry of the arguments. The blacksmith's hammer hung at his side, occasionally catching the slanting sunlight with a dull gleam that spoke of battles survived and metal reshaped by will and fire.

"Bah! They refuse to see what's right before them," he said, his voice pitched low but intense enough that a pair of sparrows roosting nearby took startled flight. "The Mirrathar signature on those mirror-shards was unmistakable. Bran verified it three different ways." He stopped his pacing abruptly and set his hammer down on a flat stone with enough force that the sound resonated against the garden walls. "And Vaelis—did you notice how her words seemed to carry an echo? As if someone else spoke just behind her voice? I swear that woman is the mirror mage!"

Below their sanctuary, Threadmere continued its perpetual commerce, untroubled by the machinations of those who sought to reshape its destiny. The clang of a smith's hammer rose from the forges in the lower districts, a familiar rhythm that might have calmed Edrin under different circumstances. Merchant ships with striped sails unloaded their wares at the crowded docks, each vessel flying pennants with its trading house and origin port. The scents of roasting chestnuts and salted fish drifted upward on thermal currents, mingling with the garden's herbal fragrances to create a perfume unique to this particular intersection of place and time. Street criers called the mid-afternoon prices for grain, cloth, and silver, their voices carrying the distinctive melodic patterns, allowing their words to be heard above the general din.

Lysera had been standing quietly near an ancient statue of Auron, her sight-veils now wrapped around her wrists like delicate bracelets. She watched Edrin's agitated movement with eyes that held understanding rather than judgment. When he paused in his pacing, she approached with quiet grace, her steps making almost no sound against the flagstones. Without hesitation, she placed her cool palm against his flushed cheek, the simple gesture bridging the gap between his anger and her calm.

"We'll find another way," she assured him, her voice carrying the same certainty she used when reading vision-patterns in the Loom. Her touch lingered, thumb brushing lightly across his cheekbone in a caress that spoke to something beyond their shared mission. "The Council's blindness doesn't change what's true."

Edrin leaned almost imperceptibly into her touch. His hand rose to cover hers, calloused fingers enfolding her slender ones with careful strength. "I know," he admitted, his voice softening. "But time works against us. Whatever mirror witch is planning, she's had days to advance it while we chased her shadow to Red Kiln and back."

A stray cat with patchy orange fur darted across the garden, pausing briefly to investigate the base of a weathered statue before continuing on its inscrutable mission. Its sudden appearance and equally abrupt departure served as a welcome interruption to the intensity of the moment. Lysera smiled faintly, turning to watch the creature vanish over the garden wall.

"Cats always know where they're going," she observed. "Even when it seems random to everyone else." She gently tugged Edrin's hand, leading him toward a stone bench positioned for the best view of both garden and city. "Perhaps we should trust our instincts as they do theirs."

They sat together on the bench, shoulders touching in companionable silence as they watched the city below transition from afternoon industry toward evening's slower rhythms. Shop owners in the tailor's district began rolling up the bright fabric samples displayed outside their storefronts. Children who had been playing complicated hopping games between chalk lines drawn on the cobblestones were called home for evening meals, their reluctant protests carrying faintly up to the garden. A lamplighter made his rounds, his long pole touching wick after wick until a trail of gentle illumination marked his passage.

"Do you ever think about after?" Edrin asked suddenly, the question emerging as if from some deep well of thought he rarely allowed himself to visit. His eyes remained fixed on the city below, perhaps finding it easier to speak of vulnerable things without direct eye contact. "After the mirror witch, after the Loom-surges, after all of this?"

Lysera's hand found his, their fingers intertwining with familiar ease. "Sometimes," she admitted. "Though it feels almost like tempting fate to imagine it too vividly."

Edrin nodded, understanding precisely what she meant. Hope was a luxury that had proven dangerous in their experience, yet its absence left a void that nothing else could quite fill. "Sometimes I imagine a small cottage," he said, the words emerging quietly, as if confessing something both precious and fragile. "Far from councils and gods and Loom-surges. Just a simple place with a forge off to the side, where I could make things that help rather than harm. Tools for building, not weapons for fighting."

The confession hung in the air between them, vulnerable in its ordinariness. It wasn't a grand vision of heroism or power—just a simple wish for peace and purpose after so much chaos and conflict. Lysera's fingers tightened around his, her touch conveying understanding without judgment.

"With a garden," she added after a moment, her voice softening as she contributed to this shared dream. "Not grand or elaborate, just practical herbs and vegetables mixed with flowers that serve no purpose except beauty." A genuine smile warmed her features as she allowed herself to elaborate. "Gloryvines climbing the walls, catching the first light. Sagum and Lorcha growing near the kitchen door. Perhaps a fruit tree that we'd watch grow taller each year."

Edrin turned to look at her now, something unguarded in his expression. "And perhaps children someday," he said, the words carrying both question and hope. "Children who would know peace as their birthright instead of a distant dream."

The admission transformed something between them. Lysera's eyes, usually so focused on distant visions, now saw only him, present and real beside her. "I would like that," she said simply, the words carrying the weight of promise.

Their fingers remained intertwined as they sat together, shoulders touching, the shared vision of that simple future that they might someday grasp. Below them, Threadmere continued its evening transition—carriage wheels rumbling over cobblestones, silk gowns rustling in courtyards as the wealthy emerged for evening entertainments, tavern doors opening to release brief bursts of laughter and music before closing again.

For these few precious moments, they allowed themselves to exist solely for each other rather than for the quest that had brought them together. The weight of responsibility, never fully absent, receded just enough to make space for something equally powerful but infinitely more gentle to flourish in its place.

#

The shadows at the garden's edge rippled like disturbed water, a subtle movement that defied the stillness of the late afternoon air. Neither Edrin nor Lysera noticed this anomaly at first, absorbed as they were in their shared vision of a peaceful future. Then came a faint metallic sound—the soft chime of a buckle against stone—followed by the sensation of presence, that indefinable awareness of being observed which prickles the skin before conscious recognition occurs. The shadows parted with deliberate theatricality, and Theryn stepped through as if emerging from a doorway only he could perceive. He moved with the fluid grace of someone to whom physical laws were mere suggestions, his green-gold eyes catching the fading sunlight in a way that made them appear to glow from within, as a predator's gaze reflected in firelight.

"Planning domestic bliss while the world burns?" he quipped, perched with casual elegance atop the garden wall where no footholds were apparent. His leather vest creaked softly as he settled. "How delightfully mortal of you."

Above him, a pennant from one of the city's watch towers fluttered in a breeze, touching nothing else in the garden, its blue and silver fabric snapping once before falling still again. The sharp sound punctuated Theryn's words like deliberate emphasis, though whether he had somehow orchestrated this coincidence remained one of many mysteries that surrounded him.

Edrin's body tensed, the peaceful openness of moments before replaced by defensive alertness. His hand dropped from Lysera's, instinctively moving toward his hammer before conscious thought intervened. The transition from vulnerability to guardedness happened in the space of a single heartbeat.

"Perhaps we wouldn't need dreams of escape if certain individuals announced their presence like normal people instead of lurking in shadows," Edrin retorted, his voice carrying an edge that belied the banality of his words. Deeper currents flowed beneath this exchange. Edrin's frustration at having a rare moment of peace intruded upon, his wariness of Theryn's unpredictable nature.

Theryn raised a placating hand, the leather of his glove brushing against the ancient stone of the wall. The gesture was both conciliatory and dismissive, somehow managing to acknowledge Edrin's irritation while simultaneously suggesting its irrelevance. "Peace, hammer-wielder," he said, his tone softening though the mischief never quite left his eyes. "Even in darkness, seeds must be planted."

The unexpected depth of this statement hung in the air between them, transforming what had begun as an intrusion into something more meaningful. From the streets below, a hawker's final cry drifted upward—the words fading into the background hum of the city. A cart rattled over cobblestones, its wooden wheels striking a loose stone that sent a clear, bell-like note echoing up to their garden perch. These ordinary sounds provided a counterpoint to the extraordinary nature of their meeting, reminding them of the world that continued its rhythms regardless of their personal dramas.

Lysera studied Theryn with quiet intensity, her gaze penetrating in a way that suggested she saw more than his physical presence. "You didn't come here merely to comment on our conversation," she observed, neither accusation nor question but a simple statement of perceived fact. "What brings you to our peaceful alcove, Theryn?"

Something flickered across his features—respect, perhaps, for her directness, or amusement at being so readily understood. With theatrical suddenness, he produced a small leather pouch from somewhere within his vest, the movement so swift it seemed almost magical. The pouch made a distinctive sound as he tossed it toward Lysera—not the dull thud of coin but the delicate chime of glass pieces striking one another with musical precision.

Lysera caught the pouch with reflexive grace, her fingers closing around it as if she had expected this exchange. The contents shifted against her palm, producing another series of crystalline notes.

"Clarity stones," Theryn explained, his posture relaxing slightly as if the transfer of this object had been his primary purpose. "For when visions blur."

A gust of wind swept through the garden, stirring rose petals at their feet and carrying the scent of rain still far beyond the horizon. The petals danced briefly before settling in new configurations that resembled glyphs of revelation and insight. Lysera opened the pouch carefully, tipping a few of its contents into her palm. Small polished stones caught the fading light, their surfaces neither fully transparent nor completely opaque but existing in some liminal state between. Each stone contained what appeared to be a suspended droplet of liquid light at its center, the droplet moving independently of the stone itself when tilted.

"These are rare," Lysera whispered, genuine surprise coloring her voice. "Formed only where ley lines cross beneath underground springs. The Archive has three in its collection, kept under seven locks." She looked up at Theryn with new questions in her eyes. "How did you come by them?"

Theryn's smile widened, revealing teeth slightly too sharp to be entirely human. "I have my methods," he replied, the deliberate vagueness suggesting stories he had no intention of sharing. "They'll help cut through illusions, particularly the mirror-kind that's been clouding Council minds lately." His expression grew momentarily serious, the shift so abrupt it was almost unsettling. "Keep them close. When the moment comes to see the truth, you'll know."

Edrin stepped closer, his earlier irritation replaced by cautious curiosity. "Why help us this way? What stake do you have in this struggle?"

For a moment, something ancient and weary flickered behind Theryn's perpetual amusement—a depth of experience that his youthful appearance belied. "Let's just say I've seen what happens when the weave unravels," he said softly. "Some patterns shouldn't be broken, even by those who think themselves above the loom's design."

Before either Edrin or Lysera could press further, Theryn rose in a fluid motion and stepped backward toward the shadow cast by a mossy column. "Use them wisely but keep them hidden until you do," he said, his form already beginning to blur at the edges as if he were becoming less substantial. "And perhaps save one of those domestic dreams for after the world doesn't end. They're quite touching, really." The last words carried the ghost of his usual mockery, but beneath it lay genuine affection.

Then, between one heartbeat and the next, he was simply gone—not hidden or concealed but absent entirely. The shadow that had swallowed him remained ordinary in appearance, revealing nothing when Edrin cautiously extended his hand into its depths.

Only the pouch of clarity stones remained as proof of Theryn's visit, their weight solid and undeniable in Lysera's palm. She closed her fingers around them, feeling a subtle vibration resonating with her sight-gift, strengthening her connection to the Loom's patterns.

"He always does that," Edrin muttered, half-annoyed and half-impressed despite himself. "Appears from nowhere, speaks in riddles, then vanishes before giving any real answers."

Lysera carefully retied the pouch and tucked it into a hidden pocket sewn into her robes. "Yet he's never steered us wrong," she observed quietly. "And these stones…" She paused, considering their significance. "These are no small gifts. Whatever game he plays, I believe he wants us to win it."

The garden settled back into stillness as twilight deepened around them, the city below gradually illuminating as lamps and hearth fires were lit against the coming night. Their moment of intimacy had been interrupted, yet something else had been given in exchange—a tool, perhaps, or a weapon against the deceptions they faced. The echo of Theryn's laughter seemed to linger in the air, a reminder that even in their darkest moments, they were not entirely alone in their struggle.

#

The Gilded Quill nestled in the crook where Archivist's Row met Scholar's Lane, its weathered façade suggesting centuries of quiet observation of Threadmere's intellectual life. Lanterns swung from wrought-iron posts outside its leaded-glass windows, their flames protected by amber panes, casting a honeyed glow across the cobblestone street. The narrow doorway, framed by shelves containing actual books—a display of wealth in a city that valued knowledge more than gold—stood slightly ajar, releasing complex aromas with stories of their own: roasted pearl barley transformed into ale, meat pies baked with imported spices, pipe smoke from exotic tobacco, and beneath it all, the distinctive scent of aged parchment and ink that clung to the tavern's regular patrons.

Bran paused before entering, his scaled fingers adjusting the strap of his leather satchel where it crossed his chest. Unlike some of his kind who avoided human establishments, Bran had long ago learned that information flowed most freely in places where alcohol loosened tongues, and professional rivalries prompted indiscretion. His scales—currently a burnished copper—would draw initial glances, certainly. But in Threadmere's academic quarter, curiosity typically outweighed prejudice, especially when the lizardfolk in question carried Archive credentials.

He pushed open the door and stepped into a room alive with intelligent minds gathering to temporarily escape their own sharpness. The space was crowded but not chaotic, filled with what appeared to be the intellectual elite of Threadmere relaxing into deliberate dishevelment after days of rigid propriety. Scribes still wore their ink-spattered cloaks, now unbuttoned and hanging loosely from shoulders hunched over drinks instead of manuscripts. Archivists with leather satchels propped carefully beside their chairs maintained the habitual gestures of their profession, handling tankards with the same reverent precision they would apply to ancient texts. Off-duty guards from the Archive's security detail occupied a corner table, their postures more relaxed than on duty but still alert, their eyes habitually scanning entrances and exit points even as they laughed over shared jokes.

The Quill's interior smelled of history—not the sterile, preserved history of the Archives, but the living, ongoing creation of it, where ideas were exchanged, alliances formed, rivalries stoked, and occasionally, significant discoveries announced after sufficient liquid courage had been consumed. Copper lamps suspended from the ceiling bathed everything in warm light, while the massive stone hearth opposite the bar radiated welcome heat against the evening's encroaching chill. Behind the bar, rows of bottles caught the light, their contents ranging from common ales to rare distillations.

The rattle of quills on parchment competed with conversation—several patrons still working despite, or perhaps fueled by, their drinks. A scribe near the hearth muttered citations to himself while cross-referencing three different scrolls, oblivious to the ale slowly soaking into his sleeve. Two young researchers with the bright blue collar-tabs of the Astronomical Guild argued passionately over star migration patterns, their hands tracing competing constellations in the air between them.

Bran approached the bar, settling onto a stool crafted specifically to accommodate non-human anatomies—a thoughtful touch that spoke to the establishment's inclusive clientele. The bartender, a woman with forearms corded from years of drawing heavy casks from the cellar, acknowledged him with a nod.

"Copper ale, if you have it," Bran requested, placing two silver coins on the polished wood surface. The brew, infused with mineral salts complementing lizardfolk palates, was becoming more common in cosmopolitan establishments.

The bartender returned moments later with a tankard whose contents glimmered with metallic highlights. "Fresh batch," she commented, "imported from the southern wetlands last week."

As Bran savored the first sip—earthy with mineral undertones—he felt rather than saw someone settle onto the stool beside him. A subtle shift in the air carried the scent of old leather, quality ink, and, beneath those professional aromas, a perfume composed of ingredients he couldn't quite identify but which suggested access to rare botanical collections.

"You're Archive, aren't you?" The woman's voice carried the precise diction of formal education tempered by confident authority. "That copper-scale pattern is distinctive—you must be Silvertail, the comparative glyph analyst."

Bran turned to face a woman whose appearance suggested scholarly credibility without ostentation. Her attire—deep blue with understated silver embroidery at the cuffs—placed her in the upper echelons of academic hierarchy without specifying her exact field. Her hair was styled in the practical braided coil favored by those who spent hours bent over research materials, and reading spectacles hung from a silver chain around her neck, though she wasn't currently wearing them. Most notably, her fingers bore the distinctive calluses of someone who regularly handled both quill and ancient binding materials.

"My reputation precedes me, it seems," Bran replied, setting his tankard down with deliberate care. "Though you have me at a disadvantage."

The woman smiled, the expression warming features that might otherwise have seemed severe. "Meritia Voss, Restricted

Collections," she offered, extending a hand in the scholar's greeting—fingers splayed to show ink-stains that served as badges of academic authenticity. "Your work on comparative symbolism in pre-Concord glyphs crossed my desk last month. Brilliant analysis of the Emberfield variations."

The conversation flowed with surprising ease, their shared professional vocabulary allowing them to exchange complex ideas with an impenetrable efficiency. Meritia spoke with authority about ancient texts that few had access to, her knowledge appearing both broad and remarkably specific.

"The mirror-glyph variations you mentioned in your third appendix," she said, leaning closer as if sharing confidential information, "there's a related manuscript collection that never made it into the main Archive catalogs. Family records from one of the old Mirrathar lineages, donated under sealed conditions."

Bran's interest sharpened visibly, his scales brightening to a more attentive copper. "That would fill significant gaps in the symbolic progression I've been tracking," he acknowledged, scholarly excitement momentarily overriding caution.

Meritia signaled the bartender, ordering a second round with a gesture suggesting familiarity. "This one's my contribution to academic collaboration," she said with a conspiratorial smile. When the fresh tankards arrived, she raised hers in a toast. "To illuminating shadowed knowledge."

The copper ale in the second tankard tasted subtly different—a bit sweeter, with an underlying note Bran couldn't immediately identify. He attributed the variation to a different cask or perhaps additives unique to this establishment. They continued their discussion of restricted archives and mirror-symbolism as he drank, Meritia gradually steering the conversation toward his specific findings regarding the illusionist's mirror-magic.

It wasn't until his vision began to blur around the edges that Bran realized something was wrong. Words suddenly required unusual effort to form, and his normally precise recall became frustratingly fragmented. His hand moved toward his satchel in instinctive protection of his research materials, but the limb seemed disconnected from his intentions, responding sluggishly if at all.

"You're looking rather pale," Meritia observed, her tone concerned, though her eyes remained calculating. "Perhaps some air would help?" Her hand closed around his upper arm with surprising strength, guiding him toward a side exit he hadn't noticed earlier.

Bran attempted to resist, to call for assistance, but his tongue felt swollen and unresponsive. The room tilted and swirled around him as consciousness began to slip away. His last coherent thought was recognition of his own carelessness—a researcher who had forgotten to properly research his companion.

Darkness claimed him between one labored breath and the next.

When awareness returned, it arrived in fragments—cold dampness seeping through his clothing, the distant clamor of late-night merchants rolling carts across cobblestones, the acrid smell of refuse mingling with fog that carried the river's murky scent. Bran's head throbbed with dull persistence. When he attempted to move, he discovered his wrists and ankles bound with thin wire that bit painfully into his scales when he struggled.

He lay in what appeared to be a service alley behind a row of shops, the narrow passage barely wide enough for delivery carts to navigate. Fog had settled into the space, transforming nearby lanterns into hazy spheres of illumination that provided more atmosphere than actual light. The buildings on either side loomed like canyon walls, their upper stories disappearing into mist. From somewhere beyond his limited view came shouts of dockhands securing the last vessels before full night fell, their voices distorted by distance and fog into something almost musical.

Bran twisted against his bonds, wincing as the wire cut deeper into his scales. His clothing had been searched with methodical thoroughness, pockets turned inside out. More alarmingly, his satchel was gone—along with the research notes on mirror-magic, the catalog of Mirrathar glyphs, and most importantly, the fragment of mirror-shard he'd collected from Red Kiln for further analysis. His coin purse had vanished, though that loss seemed trivial compared to the research materials.

"The scholar outfoxed by the fox," came a familiar voice from above. Theryn crouched on a narrow ledge that ran along the building's wall, his form materializing from shadow with characteristic dramatic flair. "Perhaps next time trust less in fine clothes and more in fine instincts." Despite the mocking words, there was genuine concern beneath his usual sardonic tone.

With a graceful leap that momentarily defied gravity, Theryn landed beside Bran, already producing a slender blade from his sleeve. The knife flashed once in the dim light as he sliced through the wire bindings with surgical precision, careful not to damage Bran's scales further.

"Theryn," Bran croaked, his throat still dry from whatever drug had been administered. "How did you…"

"Find you?" Theryn finished for him, offering a hand to help Bran to his sitting position. "Let's just say I keep an eye on interesting people, and you, my scaled friend, have become exceptionally interesting of late." He glanced toward the alley's entrance, his heightened senses clearly detecting something beyond normal perception. "We should move soon. Whoever arranged your… academic downfall might return to ensure their work was thorough."

Bran managed to stand, though the alley continued to tilt alarmingly around him. "Who robbed me?" he demanded, steadying himself against the damp wall. "That woman, Meritia, claimed to have access to restricted texts about mirror-magic."

Theryn's expression shifted to something more serious, the perpetual amusement in his eyes briefly replaced by calculating intelligence. "Someone who knows what you seek," he answered, the deliberate vagueness clearly intentional. "Which means you're asking the right questions." He paused, head tilting slightly as if listening to sounds beyond Bran's perception. "The wrong people are becoming very interested in your research, which suggests you're getting uncomfortably close to something they don't want found."

Before Bran could press for more specific information, Theryn helped him to his feet, supporting his weight with surprising strength. "Time to disappear," he murmured, guiding Bran toward a narrow passage between buildings that seemed barely wide enough to accommodate them. “I know safe places where you can recover while we determine exactly what they were after, and why it matters enough to drug an Archive scholar in full view of Threadmere’s intellectual elite.”

As they moved into deeper shadow, Theryn’s form seemed to blur at the edges, as if he were pulling the darkness around them like a cloak. The alley behind them remained empty save for the lingering smell of spilled ale and the faint echo of Theryn’s laughter—a sound that somehow carried both warning and reassurance in its enigmatic tones.

CHAPTER ELEVEN: MISTHOLD

Chapter Eleven

Misthold – Evening (Umbersday, 1 Frostwake, TT 1215)

The village of Misthold appeared first as a smudge of smoke against the pale horizon, then as pinpricks of orange flame dancing between thatched rooftops. As the companions crested the final hill on the eastern road, the full horror revealed itself. Phantom shapes—too angular to be human—darted between cottages while villagers fled in ragged groups, their screams carrying clearly across the valley floor. Mirror-born illusions, yet with enough substance to rend and burn as they passed through solid matter.

"Malrik," Bran whispered, his scales darkening to a deep bronze as recognition dawned. "The fragmentation patterns, the half-substantive nature—these bear his signature." His clawed finger traced familiar symbols in the air, matching the faint geometric halos surrounding each phantom.

Lysera's sight-veils trembled at her wrists as she unwound them, allowing the enchanted fabric to float before her eyes. Through their translucent weave, the illusions' true nature became apparent—not random terrors but carefully orchestrated attacks to herd the villagers toward the northern edge of the settlement. "He's corralling them," she said, voice tight with controlled anger. "For what purpose, I can't yet see."

Theryn was already moving, sliding from his mount with fluid grace. "Well, I don't plan to wait around for the answer," he called over his shoulder, unsheathing twin daggers that caught the afternoon sun. "Coming, Ashwood, or shall I claim all the glory?"

Edrin dismounted without responding to Theryn's needling, his movements heavier and less fluid. The Keylie Path journey had taken its toll weeks ago, and though his body had mostly recovered, a bone-deep weariness persisted, flaring whenever he pushed himself too hard. Now was not the time for caution, however. His hand found his father's hammer at his belt, the familiar weight both comfort and promise as he surveyed the chaos below.

"Together," he said simply, then broke into a run down the hillside.

The village square churned with panicked movement—women clutching children to their breasts, men wielding farming implements against enemies that slipped through solid matter, and elders stumbling as they tried to flee. And among them, the mirror phantoms moved with terrible purpose, their forms shimmering with reflected light. They left strange after-images in their wake, trails of color and shape persisting long after they should have faded.

Edrin crashed into this chaos like a stone through ice, hammer already swinging in a wide arc. The first phantom to cross his path shattered into fragments of light and shadow, its dissolution accompanied by a sound like breaking glass. Another lunged at him from behind, and he pivoted to meet it, hammer connecting with what should have been an insubstantial illusion but instead felt like striking crystallized air. The phantom fractured but didn't disappear, reforming partially as it retreated.

"They regenerate!" he shouted to his companions. "Full strikes only, with intent behind them!"

A farm woman ran past, her apron smoldering where a phantom had brushed against her. Edrin reached out to steady her, directing her toward the southern path where Lysera was organizing an evacuation route. Another phantom lunged, and his hammer swept through it with such force that the metal head hummed with vibration. This time, the illusion disintegrated completely, its light scattering like disturbed fireflies before winking out.

Sweat dripped from Edrin's brow, stinging his eyes and blurring his vision. The hammer felt heavier with each swing, his arms burning with the effort of striking with the necessary force. He pushed deeper into the village, toward a cluster of children huddled against the wall of the communal granary, their eyes wide with terror as three phantoms converged on them from different directions.

"Here!" Edrin shouted, deliberately drawing the phantoms' attention. "Face someone who can fight back!"

Two of the spectral forms turned toward him, their features—if they could be called such—rippling like reflections in disturbed water. They moved with unnatural speed, covering the ground between them in fewer steps than should have been possible. Edrin braced himself, hammer raised, muscles coiled for the impact.

The first phantom reached him, its arm extending into something like a blade of frozen flame. Edrin's hammer swung to meet it, but the phantom was faster than he anticipated. It slipped partially past his defense, the flame-blade slicing across his side just below his ribs. Pain erupted along his flank, hot and immediate, as the phantom's quasi-real form made contact. Blood welled through his tunic, spreading in a dark stain that clung to the fabric.

Edrin staggered but didn't fall. His return strike caught the phantom at what might have been its shoulder, fragmenting it into dissolving light. The second was already upon him, however, and his parry came a fraction too slow. He felt rather than saw its touch—cold where the first had been burning—across his forearm, leaving a trail of frost-crystallized blood in its wake.

"The children," he gasped, turning to see that the third phantom had not been distracted by his challenge. It loomed over the huddled group, its form expanding like smoke caught in a draft. A young boy, no more than seven or eight, broke from the group in blind panic, stumbling directly into the phantom's path.

Edrin moved without thought, pain, and exhaustion momentarily forgotten. Three long strides brought him between the child and the phantom. His hammer came up, but too slowly—the wound in his side screamed in protest, muscles failing to respond with their usual strength. The phantom's arm became a whip of scintillating energy coiling to strike.

Edrin saw he couldn't complete his swing in time. Instead, he twisted, placing his body as a shield between the phantom and the boy. He felt the whip connect with his back, a line of fire that burned through leather and cloth to the skin beneath. The impact drove him to one knee, but his arm shot out, sweeping the child behind him.

"Run," he commanded through gritted teeth, his vision swimming with black spots. "Straight for the hill. Don't look back."

The boy fled, bare feet slapping against packed earth. Edrin forced himself upright, turning to face the phantom, bleeding from his two wounds. His hammer felt impossibly heavy, his arms leaden with fatigue. The pain in his side flared with each breath, suggesting the wound was deeper than he'd initially thought.

Yet still he stood, blood-slick fingers tightening around the hammer's haft. The phantom reformed before him. For a moment, Edrin thought he saw a woman's face overlapped where the face would have been on the phantom, but it was there and gone in a flash. Edrin's legs trembled with the effort of remaining standing, but he raised his weapon once more, refusing retreat.

Across the village square, Bran was directing a group of men armed with water and cloths to cover reflective surfaces—windows, rain barrels, polished metal. His voice carried clear instructions about breaking the phantoms' anchors to the physical world. Theryn darted between buildings, his blades flashing as he struck with precision at points where the illusions seemed most substantial. Lysera had established a safe corridor along the southern path, her sight-veils creating a boundary that the phantoms couldn't easily cross.

They were making progress, saving lives, but more phantoms continued to materialize throughout the village. And at the northern edge, barely visible through the chaos, a hooded figure stood watching—hands raised in elaborate gestures directing the illusions' movements.

Edrin saw all this in fragments as he fought to remain conscious, blood loss and exertion taking their toll. The phantom before him grew larger, feeding perhaps on his weakening state. His hammer came up one more time, the movement sending fresh agony through his wounded side.

"Come then," he whispered to the phantom, blood trickling from the corner of his mouth. "Let's see which of us breaks first."

The illusion surged forward, and Edrin's hammer swung to meet it, his determination transcending the physical limitations of his wounded body.

#

Lysera felt the shift in the Loom's pattern before she saw it happen—a sickening lurch in the invisible weave. Her sight-veils flared with sudden brightness as they translated this disturbance into visual warning: Edrin's thread, burning bright with purpose moments before, now threatened to unspool from the greater pattern. She turned sharply toward the village center, veil-enhanced vision cutting through phantom-filled smoke to find him—on one knee, blood darkening his tunic and still somehow fighting, each movement weaker than the last, yet refusing surrender.

"No," she whispered, the word both denial and prayer. The evacuation route she'd established along the southern path was holding; villagers streamed to safety under the protection of her boundary glyphs. She could maintain it with minimal focus now that the pattern had set.

Edrin was another matter entirely. Even at this distance, she could see the pallor beneath his sun-bronzed skin, the unnatural stiffness in his movements. His hammer swung in an arc that should have been powerful but instead wavered mid-strike, his strength failing as blood continued to seep through his tunic in an expanding stain.

Lysera abandoned her post without hesitation, fingers tracing quick reinforcement glyphs to maintain the evacuation boundary. "Keep them moving," she instructed the village headwoman who had been assisting her. "The path will hold for another quarter-hour at least."

Then she was running, sight-veils streaming behind her like ribbons caught in a gale. She navigated the chaos with preternatural awareness, her enhanced vision revealing the safest path through the mayhem. A phantom lunged at her from behind a rain barrel, but she was already sliding beneath its reach. Another appeared directly in her path, and she threw a handful of powder from her belt pouch—powdered salt mixed with crushed mirror-lotus—disrupting its form long enough for her to pass.

Edrin's hammer finally struck true, shattering the phantom before him into dissolving light. The effort cost him his remaining strength. The weapon slipped from nerveless fingers, thudding against the packed earth as he swayed, his face ashen beneath streaks of soot and blood. His eyes rolled back, consciousness fleeing, body beginning the slow collapse toward the ground.

Lysera reached him just as he fell, sliding to her knees to catch him against her shoulder. The impact drove the breath from her lungs, his solid weight nearly bearing both of them to the earth. Blood immediately soaked into her robes where his wounded side pressed against her.

"Edrin!" His name tore from her throat as she lowered him, cradling his head while her free hand pressed against the worst wound on his side. Blood welled between her fingers, hot and frighteningly abundant. "Edrin, stay with me. Focus on my voice."

His eyes flickered open at her command, pupils dilated with shock and pain. Recognition dawned in their depths, followed by an attempt at a reassuring smile. "The boy?" he managed, voice barely audible.

"Safe," she assured him, though she had no way of knowing. It was what he needed to hear. "Now be still. Let me work."

All around them, the chaos continued—screams and shouts, the crystalline sound of phantoms shattering, the crackle of small fires catching in thatch and timber. Yet within the small space they occupied, time seemed to slow, the world's edges blurring as Lysera gathered her focus. Her hands began to emit a subtle silvery glow, not the brilliant radiance of a true healer but the gentler light of one who understood the body's patterns well enough.

"This will hurt," she warned, no time for gentleness. She placed both glowing palms directly against his wound, feeling the torn flesh and damaged vessels beneath. Blood continued to flow between her fingers, but she ignored it, concentrating instead on the pattern beneath—the intricate weave of muscle, vein, and sinew composing his physical form.

Edrin's body went rigid beneath her touch, a grunt of pain escaping through clenched teeth. His hand shot up to grasp her wrist, not to push her away but to anchor himself against the fresh wave of agony her healing touch had unleashed.

"Hold still," she instructed, her voice steady despite the panic fluttering in her chest. "I'm not a healer, not truly, but I can stabilize you."

Her fingers moved in precise patterns over his wound, leaving trails of silver light. She wasn't closing the wound—that was beyond her skill—but rather stemming the blood flow, reinforcing weakened vessels, buying time until proper healing could be found. A temporary measure, but perhaps enough to save his life.

A healing glyph formed beneath her palms—not drawn or carved but manifested through pure intent and desperate need. Unlike the complex glyphs used by trained healers, this was simpler, almost primal in its geometry. Silver light seeped into torn flesh, temporarily binding what had been separated, reinforcing what had been weakened.

Edrin's breathing, which had been shallow and rapid, gradually steadied. The hand gripping her wrist relaxed slightly, though he didn't release her. Color returned to his face by increments, the deathly pallor giving way to merely exhausted. His eyes, which had been unfocused with pain, sharpened with returning awareness.

Their gazes met as the healing glyph completed its work, its light dimming as it sank fully into his flesh. Something passed between them in that moment—recognition of how close death had come, gratitude for its temporary postponement, and something deeper that neither had fully acknowledged.

"You shouldn't have risked yourself," he murmured, though his eyes conveyed the opposite sentiment.

"As if I had a choice," Lysera replied, her voice softer than intended. Her hand remained on his side, no longer glowing but maintaining connection, feeling his heartbeat steadying beneath her palm.

The chaos around them seemed to recede further, awareness narrowing to this connection amid the greater storm. Edrin's hand moved from her wrist to her face, fingers trembling slightly as they brushed a strand of hair from her cheek. The touch lingered, warm against her skin despite the blood loss and trauma his body had endured.

Lysera found herself leaning forward, drawn by something beyond conscious thought. The space between them contracted, breath mingling in the narrowing gap. His eyes—dark as rain-soaked earth—held questions and certainties in equal measure. Her own gaze dropped briefly to his lips, parted slightly with each careful breath.

Just as the space between them had nearly vanished, a massive crash erupted nearby. One of the larger phantoms slammed into the village well, shattering the wooden structure and sending splintered debris flying in all directions. A jagged piece of timber struck the ground inches from them, jolting them back to the reality of their circumstances.

The moment fractured like the well itself, an intimate connection giving way to immediate danger. Lysera pulled back, though her hand remained on Edrin's wound, monitoring the temporary healing she'd established. Conflict wrote itself across her features—duty to their mission, battling with the desire to remain in that suspended moment.

"Can you stand?" she asked, practicality reasserting itself, though her voice carried the echo of what had nearly transpired between them.

Edrin nodded, determination hardening his features, though pain still shadowed his eyes. "The bleeding's stopped," he observed, looking down at his side where the wound remained open but no longer pumped his life onto the ground. "Your work?"

"A temporary measure only," Lysera cautioned, helping him to his feet with an arm around his waist. Her touch lingered there, supporting him but also maintaining the connection that had nearly transcended the physical moments before. "The wounds need proper healing—I've merely convinced your body to hold itself together for now."

He swayed slightly as he stood, leaning into her support more heavily than he would have allowed himself under different circumstances. His arm settled across her shoulders, humbly accepting the necessity.

"It'll do," he said simply, reaching down with his free hand to retrieve his fallen hammer. The movement caused him to wince, but he managed it, the weapon returning to his grip like an old friend. "Until we finish this."

Lysera's hand tightened on his waist, offering both support and something more complex—acknowledgment of the moment that had passed between them and the promise that such moments might come again, should they survive the day. "Until we finish this," she agreed.

Together they turned to face the continuing battle, the temporarily healed wound binding them physically just as the unspoken moment bound them emotionally—both connections fragile, impermanent, yet powerful enough to sustain what needed sustaining until something more permanent could be found.

#

While Lysera tended to Edrin's wounds, Bran stalked the perimeter of the village square. His reptilian gaze, behind his glyphed glasses, dissected the phantoms' movements with scientific precision. Each illusion followed subtle patterns—not random terrors but coordinated attacks. His clawed fingers traced observations in a small leather-bound notebook, mapping trajectories, documenting how certain phantoms phased through solid objects while others left scorch marks in their wake. Behind this apparent chaos lay a mathematical certainty, a structure waiting to be decoded. And in structure, Bran knew, always lay weakness.

"They return to reflective surfaces," he muttered, scales shifting to a contemplative copper as he observed a particularly aggressive phantom. It had harried a group of villagers toward a water trough, then seemed to gain strength after passing over the still surface. "Not random at all—they're anchored to mirrors, to reflection."

Across the village, Theryn climbed atop an overturned cart, his lithe form silhouetted against smoke-filled sky. Unlike Bran's methodical study, Theryn had instinctively grasped the immediate need for organization amid panic. His voice cut through the din with surprising authority, all trace of his usual sardonic tone replaced by commanding clarity.

"Form groups of three!" he called to the scattered villagers who had not yet fled. "One to distract, two to attack from behind! They can't track multiple targets!" His hands gestured sharply, directing impromptu squads toward strategic positions. "The smith—bring cloths, bring covers! Anything reflective must be hidden!"

Villagers who moments before had run panicked now rallied, finding purpose in Theryn's commands. A burly blacksmith distributed damp cloths to runners who draped them over windows and looking glasses. Two teenage boys carried buckets of mud, splattering it across rain barrels and polished metal tools. Wherever reflections disappeared, the phantoms lost cohesion, their forms wavering like candle flames in the wind.

Bran's eyes widened as understanding crystallized. He snapped his notebook shut and sprinted toward Theryn. "The mirror-sigil is embedded in the phantoms," he called out, staff tracing complex symbols in the air that momentarily disrupted a phantom's advance. "They're anchored to reflective surfaces!"

Theryn nodded sharply, immediately incorporating this insight into his strategy. He vaulted higher onto a stack of crates, whistling with ear-piercing shrillness that drew all eyes. "Water barrels, windows, cover them all!" he shouted to the villagers. "Anything that throws back light gives them strength! Mud it, clothe it, break it if you must!"

With renewed purpose, the defensive efforts gained momentum. Phantoms that had terrified the village now found their movements restricted as their anchors disappeared one by one. Those that remained grew more desperate, their attacks more frenzied but increasingly unfocused.

Bran wiped sweat from his scaled brow, satisfaction evident in his posture despite the exertion. "Mathematical certainty," he murmured to himself. "Even illusions obey laws, if one knows how to read them."

Across the village square, Lysera helped Edrin to his feet. Though temporarily stabilized, his wounds remained serious, requiring proper healing that lay beyond her abilities. His face bore the gray pallor of significant blood loss, yet determination hardened his jaw as he gripped his hammer, refusing to yield to weakness while the threat persisted.

The four companions converged at the village center, each approaching from different vectors, forming, perhaps not coincidentally, a perfect square. In that geometrically significant moment, they witnessed a new wave of phantoms coalescing at the northern edge of the village—larger, more substantial, their forms bearing greater resemblance to actual beings rather than abstract shapes.

"He's changing tactics," Bran observed, staff raised defensively. "These new constructs draw power from fear directly, not just from reflections."

"Fear feeds them, then we starve them," Theryn replied, twirling his blades with deceptive casualness that masked their deadly precision.

Lysera's hand remained at Edrin's waist, supporting him while her sight-veils trembled with the strain of continued use.

"Edrin should retreat," she urged, concern evident in her voice. "The temporary binding won't hold through another fight."

Edrin's response was to straighten despite the pain that flashed across his features. "No retreat," he said simply. His free hand found hers where it rested against his side, blood-crusted fingers entwining with her clean ones.

Something shifted in that touch—a connection deeper than flesh, an alignment of purpose and protection. The moment that had begun during her healing, interrupted by battle but not severed, reasserted itself with surprising intensity. Edrin felt it first as warmth spreading from where their hands joined, then as a strange lightness that belied his injuries. It wasn't healing—his wounds remained—but something else entirely.

"Stay close to me," he commanded, though his voice held no arrogance, only certainty born of sudden understanding.

The first line of phantoms surged forward, their forms rippling with malicious intent. Edrin planted his feet, hammer raised, despite the tremor in his arms that betrayed his weakened state. The ground beneath him seemed to pulse once in sympathy, then again with greater intensity. Lines of light spread outward from where he stood, hairline fractures in reality that leaked golden radiance.

The pattern began as a simple circle surrounding the four companions, then rapidly elaborated itself—vines of light spiraling outward in curves and whorls that obeyed no obvious mathematical principle yet possessed undeniable harmony. Where the light touched earth, it transformed the ordinary dirt into something lustrous, as if microscopic particles of crystal had always existed in the soil, merely waiting for this illumination to reveal themselves.

"What is this?" Theryn asked, uncharacteristic wonder replacing his usual cynicism as the pattern enclosed them.

Edrin shook his head, equally bewildered yet somehow certain of the protection manifesting through him. "I don't know how, but it feels… right."

The pattern continued to elaborate itself, the spirals connecting into a complex glyph pulsing with Edrin's heartbeat. As the leading phantoms reached their outer boundary, they recoiled as if striking an invisible wall. The glyph flared brighter at each point of contact, repelling the illusions with quiet but implacable force.

Bran's eyes widened with recognition, his scales flushing to a deeper hue as he observed the glyph's specific configuration. The central shape, the particular angle of the spirals, the way the energy flowed through the pattern—all spoke to a particular category of protection glyph, one born not of study or training but of instinctive emotion. He remained silent, however, merely nodding to himself as understanding dawned.

"It manifested as a shield," Lysera whispered, her sight-veils revealing aspects of the glyph invisible to normal perception. "But how—?"

"Questions later," Edrin interrupted, newfound strength entering his voice. "We move forward now. Together."

As one, they advanced, the glyph-shield moving with them like a living dome of energy. Wherever they approached, phantoms scattered or dissipated upon contact with the barrier.

Villagers who had been cowering in doorways emerged, awestruck by the golden light.

"Stay within the light," Lysera called to them, gesturing families closer. As more villagers entered the shield's protection, something remarkable occurred. The glyph expanded, its patterns grew more elaborate rather than stretching thin.

They moved methodically through the village, the glyph-shield clearing a path of safety. Phantoms that approached the barrier shattered into fragments of light, their dissolution accompanied by sounds like distant windchimes. Those who attempted to phase through solid objects to bypass the shield found themselves repelled with even greater force.

From the northern edge of the village, a hooded figure observed the unexpected development with evident displeasure. The figure's hands, previously weaving patterns directing the phantoms' movements, now dropped to its sides. The remaining illusions wavered, losing coherence as their creator's concentration broke.

"Malrik," Bran confirmed, his keen sight picking out details others might miss—the distinctive embroidery on the hood, the particular stance that matched descriptions from witnesses they had interviewed. "He's retreating."

Indeed, the figure had begun backing away, abandoning its position as the glyph-shield continued its inexorable advance. A final gesture sent the remaining phantoms surging forward in what appeared to be a desperate attempt to breach the barrier, but they shattered against the golden light. Then the figure turned and fled into the forest beyond the village boundary, disappearing among the trees with unnatural speed.

The phantoms, bereft of their creator's will, dissolved one by one until only ordinary smoke and fire remained to be dealt with. Villagers emerged fully from hiding, approaching the golden dome with cautious reverence. Children reached out to touch its surface, laughing with delight when their fingers passed through without resistance, the light responding to their innocence with gentle ripples.

"It recognizes intent," Bran observed quietly, scientific curiosity momentarily overcoming his discretion. "Fascinating."

As the immediate threat receded, the glyph's light began to dim, its elaborate patterns simplifying until only the central motif remained, two paths intersecting and continuing onward as one. Then even this faded, sinking into the earth like water absorbed by parched soil, leaving behind only a faint luminescence that would not fully disappear until sunset.

Edrin swayed as the energy that had momentarily sustained him ebbed away, his wounds reasserting their claim on his strength. Lysera moved to support him, her arm around his waist, both necessary and desired. Theryn and Bran joined them, forming a tight circle amid the village they had saved.

"The village is secured," Theryn reported, uncharacteristically formal. "Though damaged, most structures remain intact. No villagers killed, though several bear injuries." His eyes flicked to Edrin's blood-stained tunic. "Some more severe than others."

"And Malrik?" Edrin asked, voice rough with pain but unwavering.

"Gone, but not victorious," Bran replied, tucking his notebook away with careful precision. "He tested something here today—perhaps the range of his illusions, perhaps their ability to manifest physical effects. Either way, he failed."

Lysera's free hand reached for her sight-veils, which had dimmed to their normal state. "He'll try again elsewhere," she said softly. "But we've learned much about his methods, and..." She glanced at the ground where the glyph had been. "And perhaps discovered new ways to counter them."

Around them, the village slowly returned to order. Fires were extinguished, wounds bandaged, and children comforted. The afternoon sun hung low on the horizon, casting long shadows stretching toward the eastern hills where their journey had begun just hours before. The air smelled of smoke and sweat and blood, but beneath those harsher notes lay the scent of turned earth and growing things—life continuing despite all attempts to disrupt it.

"We should tend to your wounds properly," Lysera told Edrin, her voice carrying both professional concern and personal tenderness. "What I did will hold for now, but..."

"But it's enough," Edrin finished for her, his hand finding hers again with purpose rather than desperation. "It's enough to continue. That's all we need."

In that simple exchange lay the echo of the glyph that had protected them—connection transmuted into protection, emotion given form through necessity and will. The four companions stood united in their victory, knowing they had gained valuable insight into their enemy's tactics even as they discovered unexpected strength within themselves.

The village was saved, though scorched and damaged, and they stood together in their hard-won triumph. Malrik had retreated, but the confrontation had only begun.

A prelude to greater challenges awaiting them on the road ahead.

CHAPTER TWELVE:
MISTHOLD

Chapter Twelve

Misthold – Morning (Commonsday, 28 Frostwake, TT 1215)

Frost crunched beneath the villagers' boots, the sound like tiny bones breaking in the crisp morning air. Winter had laid claim to Misthold with an artist's precision, etching delicate patterns across windows and painting the thatched roofs with a dusting of white. The village moved with renewed purpose after the phantom attack, as if the very act of decorating their humble homes with evergreen garlands and pale blue ribbons might ward off the memory. Children darted between adults, trailing ribbons like comets' tails, their laughter a defiant counterpoint to yesterday's screams.

Through this renewal walked Edrin Ashwood, each step a negotiation with pain. His side throbbed beneath hastily applied bandages, Lysera's temporary healing having prevented his death but not his suffering. His face remained pale despite the cold that should have flushed his cheeks, and his breathing came shallow and measured.

Lysera walked beside him, her gaze alternating between the village preparations and Edrin's face, alert for any sign that his strength might fail. Her fingers brushed his arm with deliberate casualness whenever his breath caught or his stride faltered—a touch that conveyed both support and something more intimate.

"You should be resting," she murmured as they paused beneath the arch of the village square, where two young women struggled to secure an unruly garland of pine and juniper. "That binding won't hold if you keep testing its limits."

Edrin's mouth curved into a smile that was part grimace. "I've never been good at being still," he replied, watching the women's efforts. "Besides, I prefer this pain to the one that comes from sitting alone with my thoughts."

The wind shifted, carrying the scent of pine and woodsmoke, along with fragments of conversation from villagers who moved around them like water around stones in a stream. Talk of the festival mingled with hushed recountings of yesterday's attack, terror being transformed into story even as the village transformed itself with winter decorations.

"Here, let me," Edrin said suddenly, stepping forward to reach for the troublesome garland. The movement pulled at his wound, and he couldn't suppress a wince that Lysera caught immediately.

"Stop," she commanded, her hand firmer on his arm now. "Your stubbornness will undo what little healing I managed." She turned to a young boy who had been watching them with undisguised curiosity. "You…what's your name?"

The child startled at being addressed, his eyes widening. "Tren, miss."

"Tren, I need a favor that only the quickest among you could manage," Lysera said, crouching to meet his eyes. "Do you know the town of Threadmere? Good. I need you to carry a message there, to a healer named Alira. Tell her Edrin Ashwood requires her skills in Misthold, with all haste." She tossed a few coins at the boy.

As she spoke, her eyes flicked toward Edrin's bandaged side, the worry in them visible to all. The boy nodded, excitement at the important task overtaking his fear, and darted away with the message clutched like a treasure in his mind.

"Alira has other duties," Edrin protested quietly when the boy had gone. "This isn't worth…"

"You are worth it," Lysera interrupted. The words hung in the air between them, impossible to dismiss as mere concern for a comrade. Something shifted in Edrin's expression—surprise yielding to a deeper understanding that softened the lines of pain around his eyes.

They resumed their walk, moving toward the village green where a circle of stones had been arranged for the winter festival's central fire. An elderly couple worked nearby, tying blue ribbons to branches to be cast into the flames as wishes for the coming year. Without discussion, Edrin and Lysera joined them, their hands moving with practiced efficiency despite Edrin's injury.

As they worked, their fingers occasionally met around the same ribbon or branch, lingering in these brief connections longer than necessity demanded. The elderly couple exchanged knowing glances but continued their work without comment.

"I have been thinking, again, about the after," Edrin said quietly, his voice meant only for Lysera, though the words carried deeper meaning than their simple form suggested. "After Malrik. After the Loom-threat."

Lysera's hands stilled on the ribbon she was tying. "Someday," she replied, equally soft. "After all this." Her eyes met his, and in them lay the continuation of the moment that had been interrupted in the heat of battle.

"A small place," he continued, returning to the vision they had shared in Threadmere's garden. "Nothing grand. Just somewhere quiet."

"With a garden," she added, completing their shared dream. "And perhaps..." She stopped, bashfulness touching her features despite the gravity of all they had witnessed together.

The elderly woman working beside them cackled softly, her withered hands never pausing in their deft movements. "Young love speaks in futures," she observed to no one in particular. "As if tomorrow might notice it's being planned for."

Around them, other villagers had paused in their work, drawn by the quiet intensity between the two strangers who had saved their homes. A sense of proprietary interest had developed among Misthold's people—these heroes were now, in some small way, theirs. They belonged to the village's story, and the village to theirs.

"You should tie your ribbon together," suggested a young mother, her infant sleeping against her shoulder. "For blessing."

Edrin and Lysera found themselves at the center of a small circle of expectant faces. The rhythm of the village's preparation had subtly shifted. Lysera held a length of pale blue ribbon, its ends trailing toward Edrin's outstretched hand.

"For wishes shared," the elderly woman intoned, her voice carrying the weight of tradition. "For burdens halved and joys doubled." She approached them, laying a hand atop theirs where they held the ribbon between them. "May the binding of your hands foretell the binding of your lives, and may your children's children gather beneath this same winter sky to remember how you found each other amid shadows."

The formal wedding blessing fell into a sudden silence, its implication spreading ripples through the watching villagers. Someone gasped softly. A child giggled. Edrin's face flushed despite the blood he had lost, and Lysera's eyes widened at the misunderstanding.

"We're not..." Edrin began, but found the denial stuck in his throat like a fish bone, impossible to either swallow or expel.

"What he means is..." Lysera attempted, only to discover herself equally incapable of contradicting the assumption that hummed in the air around them.

The elderly woman merely smiled, patting their joined hands before stepping back. "The heart knows before the mouth speaks," she said with the certainty of one who had witnessed this truth unfold across decades. "Tie your ribbon. Be ready to be married. Make your wish. The rest will follow in its time."

And so they did, fingers brushing as they secured the ribbon to a branch that would later be consumed by ritual fire, releasing their unspoken wish to the winter sky. Around them, the village returned to its preparations with renewed energy. This small ceremony had somehow completed a circuit, allowing joy to flow more freely through the community that had so recently known only terror.

Edrin's hand found Lysera's as they stepped away from the circle, his touch conveying what words had failed to express. For this brief moment, amid frost and evergreen and the simple rituals of village life, they allowed themselves to exist not as warriors against cosmic threats but as two people contemplating a shared future.

#

Midday light filtered through the frost-laced windows of the village elder's home, fracturing into prisms dancing across the worn floorboards like living entities. The room smelled of dried herbs hung from ceiling beams and the sweet pine smoke curling from the stone hearth, where a kettle hummed quiet promises of warmth. Edrin sat on a bench carved from a single oak trunk, his shirt carefully folded beside him, exposing the angry wound that curved from his lower ribs around to his back—a testament to yesterday's battle that pulsed beneath Lysera's temporary binding.

The door opened with a whisper of cold air, bringing with it a figure silhouetted against the bright snow outside. Alira stepped into the room with quiet grace, her healer's satchel slung across one shoulder and her iridescent wings folding neatly against her back to accommodate the low doorway. Frost clung to her auburn hair in a crown of miniature stars that began to melt almost immediately in the room's warmth, sending tiny rivulets down strands that shimmered with hidden greens and golds.

"You've been fighting phantoms without me," she said by way of greeting, her voice carrying the musical lilt characteristic of fairykin. She set her satchel on the table with practiced ease and unwound the scarf from her neck, revealing the silver pendant that hung there, its petals seeming to pulse with subtle light.

Lysera rose from where she had been sitting beside Edrin, relief evident in the softening of her shoulders. "Thank you for coming so quickly," she said, stepping back to give Alira room to work. "My binding was temporary at best. The phantom's touch left something… residual."

Alira nodded, her forest-green eyes taking in the situation with professional assessment that held no judgment. She approached Edrin, her movements deliberate and unhurried as she knelt before him. "May I?" she asked, hands hovering near but not yet touching his wound.

Edrin nodded, his jaw tight with pain he refused to voice. Alira's fingers were cool against his fevered skin as she gently probed the edges of the wound, her touch clinical yet somehow comforting. The jagged tear had begun to knit together at its edges thanks to Lysera's emergency glyph work, but the deeper damage remained—muscle improperly aligned, small vessels leaking blood into surrounding tissue, and an angry heat that spoke of brewing infection.

"The phantom's touch carries corruption," Alira explained, as much for her own process as for their benefit. "It's not just physical damage. There's a disturbance in your body's natural leyline flow." Her fingers traced the invisible energy pathways that ran beneath his skin, feeling the blockages and diversions caused by the injury.

Edrin grunted, "Now I have leylines in me? When did that happen?"

Lysera rolled her eyes. Alira straightened and opened her satchel, removing several small bottles sealed with wax and cork.

With methodical precision, she arranged them on the table, each placement deliberate. From a side pocket, she withdrew a small bundle wrapped in silk—when unfolded, it revealed a set of slender stylus-like tools made from various materials: silver, bone, crystal, and what appeared to be petrified wood.

"This will feel... Strange," she warned, selecting the crystal stylus. "But the connection must be reestablished before true healing can begin."

With the tool poised between her fingers like a painter's brush, Alira began to sketch shimmering patterns in the air around Edrin's wound. The glyphs hung suspended, luminous threads of energy that resembled script in a language too old for human memory. Each completed symbol pulsed once before settling onto Edrin's skin with a soft hiss, like water droplets on hot stone.

Edrin's breath caught as the first glyph made contact, his muscles tensing instinctively. But instead of pain, the sensation was one of profound rightness. The second glyph brought a cooling sensation spreading outward from the point of contact, soothing the angry heat of brewing infection. The third seemed to sink deeper, beyond flesh and into the very flow of his life force, restoring channels that had been diverted by trauma.

Throughout the process, Alira's face remained composed in concentration, though a fine sheen of perspiration appeared on her brow. Fairykin healing came at a cost—a temporary sharing of the patient's burden that manifested as physical stress in the healer. Yet she worked without hesitation, each glyph placed with precision that spoke of both training and innate gift.

Lysera watched with quiet fascination, recognizing in Alira's work an advanced version of her own emergency binding—a testament to what properly channeled power could accomplish. The air in the room grew thick with invisible energy.

Finally, Alira placed the last glyph—a complex spiral that settled over the center of the wound and pulsed with a rhythm matching Edrin's heartbeat. She sat back on her heels, tucking a strand of damp hair behind her ear with a hand that trembled slightly from exertion.

"There," she said softly. "The channels are restored, and your body's natural healing can now proceed correctly." She reached for a small jar of salve, applying it to the outer wound with gentle thoroughness. "The phantom's corruption is neutralized, though the memory of it may linger in dreams for a while."

The change in Edrin was immediately visible. Color returned to his face, and his breathing deepened naturally without the careful control he had maintained to minimize pain. He rolled his shoulder experimentally, finding movement fluid where before it had been restricted.

"Thank you," he said simply, the words inadequate for the gift she had given.

Alira nodded, recognizing in his tone all that remained unspoken. "The body remembers what it should be," she replied, repacking her tools with the same care with which she had arranged them. "I merely reminded you of its proper pattern."

When they emerged from the elder's home into the bright winter day, the village's preparations had progressed significantly. A great circle of evergreen boughs now defined the festival space, and blue ribbons fluttered from poles set at regular intervals around its circumference. Children worked alongside adults, their faces flushed with cold and excitement as they carried armfuls of kindling for the central fire that would burn through the night.

Edrin moved with new freedom, his stride no longer hampered by pain. The wound remained, and a scar would eventually form. However, the danger had passed, leaving behind only the body's natural healing process. Lysera walked beside him, her relief evident in the lightness of her step and the way her hand occasionally brushed his.

They joined a group of villagers weaving ceremonial ribbons into complex patterns that would later be hung from the festival poles. The work was simple but required coordination—each person holding strands that passed between many hands, creating an interconnected web that symbolized community bonds strengthened through winter's trials.

With each pass of ribbon between them, Edrin and Lysera's fingers met and lingered. They worked in companionable silence for a time, the rhythm of the weaving allowing their minds to wander toward possibilities that had seemed remote just days before.

"A small home," Edrin said quietly, continuing their earlier conversation as if no interruption had occurred. "With a forge attached, but not too close, so the noise wouldn't disturb."

Lysera smiled, her hands never faltering in their pattern. "And the garden we spoke of. Perhaps with a bench beneath a shady tree, to sit in the evenings."

"Children would need space to run," he added, his voice carrying a question beneath the statement—a future imagined but not presumed.

"And to climb," she replied, accepting the question and its implications. "Trees are important for growing spirits."

Around them, the village continued its preparations, but the center of this particular universe had narrowed to the space between their hands. The pattern they created was both part of the larger whole and uniquely theirs—strands of possibility woven into temporary form.

Laughter rippled through the group as someone told a story about last year's festival, when the headman's dog had stolen the ceremonial wreath and buried it in the snow. The sound was infectious, drawing smiles even from those who hadn't heard the tale. For this brief interlude, war's memory receded before the more immediate reality of community and celebration.

Edrin and Lysera found themselves caught in this current of simple joy, their own laughter joining the chorus. It was a moment suspended out of time—a breath between battles, a glimpse of what peace might feel like when it wasn't merely the absence of conflict but the presence of something worth preserving.

The ribbons moved between them, blue and silver catching winter light. Their hands met, separated, met again, each contact a promise unspoken but understood. And in the weaving of festival decorations, they found themselves crafting something else entirely—a future that might, with care and courage, eventually take root and grow.

#

The hidden chamber beneath Threadmere's forgotten temple breathed like a living entity, its air charged with arcane currents that flowed between the seven mirrors arranged in perfect heptagonal symmetry. Each mirror stood twice the height of a man, its frame forged from metals that had never known daylight, harvested from veins deep beneath the earth where darkness itself had weight and substance. No ordinary flame illuminated this sacred space. Instead, globes of captured twilight hung from chains of cold silver, their glow simultaneously bright enough to reveal every detail yet soft enough to preserve the chamber's essential secrecy. The walls themselves seemed to absorb sound, creating a pocket of perfect silence where even heartbeats felt like intrusions upon ancient mysteries.

The mage stood at the chamber's center, her body nearly bare save for the silver chains that wrapped her torso in geometric precision, each link positioned to amplify rather than restrict the flow of energy through her flesh. Her light brown skin caught whispers of light from the mirrored walls, transforming her into a celestial map of violet and silver constellations that shifted with each breath. The fractured planes of her eyes—like mirror shards embedded in human sockets—reflected the chamber in countless tiny perspectives, each fragment capturing a slightly different reality. Her hair, tasseled and free, floated around her head in light slivers and blue framing.

She breathed deeply, drawing the chamber's rarified air into her lungs in a controlled rhythm that gradually synchronized with the subtle pulse of energy emanating from the mirrors. "The hour approaches," she whispered, her voice carrying harmonics that caused the mirrors to vibrate in sympathy, producing tones too subtle for ordinary hearing but potent enough to realign the chamber's energies toward her purpose.

With deliberate ceremony, she approached the largest mirror—a massive sheet of polished obsidian flecked with silver inclusions that formed a natural glyph-map of the night sky. Her reflection watched her with expressions that didn't quite match her own, suggesting these weren't mere images but aspects of herself that existed in adjacent realities, peering through the thin membrane between worlds.

The sorceress's slender fingers traced complex patterns across the mirror's surface, each movement precise and purposeful. Where her skin contacted the obsidian, trails of silvery light remained—not on the surface but somehow within it, as if her touch penetrated dimensions rather than merely marking physical matter. The patterns began as simple curves and lines but rapidly evolved into intricate geometries that folded into and through themselves, creating structures that should have been mathematically impossible in three-dimensional space.

The chamber's temperature dropped precipitously as the pattern neared completion, her breath becoming visible in crystalline clouds suspended before dissolving into the hungry air. The light from the suspended globes dimmed, not from any change in their output but because the mirror ritual was consuming ambient energy to fuel its awakening.

"Mirror of shadows, eye of night, open your gaze beyond flesh's sight," the mirror witch intoned, her voice layered with subtle harmonics that caused the completed pattern to pulse with increasing intensity. "Show me the vessel, show me the man, show me the puppet who furthers my plan."

The obsidian mirror's surface rippled like disturbed water, its solid form temporarily liquefied by her command. Within its depths, colors swirled and coalesced, gradually resolving into a vision of distant reality captured with perfect fidelity—Malrik, sleeping restlessly on a simple cot in a canvas tent. Every detail was preserved: the sweat beading on his brow, the rapid movement of eyes beneath closed lids that betrayed troubled dreams, even the rhythm of his breathing as it quickened with unconscious anxiety.

The mirror witch's lips curved into a smile that contained no warmth, only satisfaction at finding her subject precisely as she had anticipated. "Dreams shape the waking man," she murmured, her hands now hovering just above the mirror's surface, fingers splayed like the legs of some exotic spider. "And I shall shape your dreams."

With methodical precision, she began to layer new elements into the vision—not changing what was real but overlaying it with what could be, what might be, what she intended to make inevitable. First came subtle alterations to Malrik's physical form: his muscles defined with greater clarity, his features sharpened into something more commanding, his very presence enlarged until he dominated the simple tent not through actual size but through implied significance.

"You were born common," she whispered to the sleeping image, "but destiny rarely respects the boundaries of birth."

Her fingers moved faster now, leaving trails of light that coalesced into a glimmering vision-glyph of extraordinary complexity. The glyph hovered above the mirror before sinking into its depths like a stone through dark water, ripples spreading outward as it merged with Malrik's dreaming form. Where it touched him, divine light bloomed—not the gentle illumination of benevolent gods but the terrible radiance of power unlimited by morality or restraint.

In the vision, Malrik's sleeping body became mantled with energy that crackled like lightning contained within flesh, its brilliance casting harsh shadows that stretched beyond the tent's confines and into the world beyond. A crown of pure force materialized above his head, not resting upon but hovering just above, suggesting power granted rather than inherited or seized.

"See yourself transformed," the mirror witch continued, her voice dropping to a hypnotic murmur that bypassed conscious thought and spoke directly to the dreaming mind. "Not merely a man but a vessel, a channel through which godhood flows into the mortal realm. Feel how they kneel before you, how they tremble at your approach, how your very presence reshapes reality according to your will."

The vision expanded further—crowds of worshippers prostrating themselves before Malrik's transfigured form, mountains reshaping themselves at his gesture, storms gathering and dispersing at his command. Throughout it all, his expression remained one of perfect satisfaction, a hunger finally sated after lifelong starvation.

With the vision firmly planted in Malrik's dreaming mind, the ritualist straightened and turned toward a smaller mirror set into the eastern wall of the chamber. Unlike the others, this mirror's frame bore glyphs of supplication and petition—channels through which mortals might address beings beyond their comprehension.

"Umbryos," she called, her voice shifting into formal cadences of ritual address. "Shadow of Transformation, Keeper of Secrets. I, mirror witch, your faithful servant, seek audience." Her hands rose, trembling from the effort of channeling sufficient power to bridge the gap between mortal and divine realms.

The mirror's surface darkened beyond mere absence of light, becoming a void that suggested infinite depth. From within that darkness came a presence—not visible in any conventional sense but perceptible as pressure against the mind, as weight upon the soul. The temperature in the chamber plummeted further, frost crystallizing along the mirror's edges despite the absence of moisture in the preserved air.

"I have served faithfully," the mirror witch continued, maintaining perfect composure despite standing nearly naked before a god's attention. "I have woven your will into the fabric of mortal events, directed their course toward the changes you desire. Now I seek elevation—a mantling ceremony that would bind a portion of your divine essence to my mortal frame, allowing me to serve as your direct instrument in the world above."

The darkness within the mirror pulsed once—acknowledgment without agreement. Then came the response, deposited directly into the watcher's consciousness: Denial. Timing inappropriate. Balance delicate. Vessel inadequate.

Had anyone else been present to witness, they might have seen a flicker of rage distort the veiled adept's composed features—a momentary break in her perfect control quickly mastered and transformed into renewed determination. The rejection burned like acid against her ambition, but she had anticipated this possibility and prepared accordingly.

"I understand, my lord," she replied, her tone betraying none of her inner fury. "If I am unsuitable as a vessel, perhaps another might serve your purpose better." With calculated hesitation, she added, "The one I have prepared—Malrik—his body grows stronger with each trial. His mind bends increasingly toward your service. Perhaps he might serve as your avatar while I continue my work from the shadows."

The darkness pulsed again, this time with interest rather than dismissal. Concepts flowed: Consideration. Potential. Requirements. Preparation necessary.

The mirror witch's lips curved once more into that cold smile, triumph carefully contained behind a mask of humble service. "I shall prepare him further," she promised, her fractured eyes gleaming with calculation. "Through dreams and trials, I will shape him into a perfect vessel for your divine will. A mortal avatar through whom your power might reshape this flawed world."

As the divine presence withdrew from the mirror, returning it to ordinary reflectivity, the mirror-worker turned back to the obsidian glass where Malrik still slept, unaware of how thoroughly his fate was being manipulated. Her finger traced one final pattern into the vision—a subtle glyph that would lodge in his subconscious like a splinter too deep to extract.

"Serve as conduit," she whispered, her voice barely audible even in the chamber's perfect silence. "Channel divine power that would break lesser men. And when you have opened the way, when the boundaries between mortal and divine have been sufficiently weakened…"

She left the thought uncompleted, but her reflection—or perhaps some aspect of herself from another reality—finished it with a smile that revealed teeth too sharp for human mouths, eyes too knowing for mortal comprehension. The plan unfolded across multiple dimensions, each layer visible only from certain perspectives, its totality comprehensible only to one whose vision had been shattered into perfect fragments.

The mirrors watched in silent witness as the practitioner completed her ritual, each reflection showing a slightly different version of events, a different outcome, a different reality—yet all converging on a single, inevitable conclusion that would elevate her far beyond the role of mere servant to a god who understood so little of her true ambition.

#

Malrik woke with violence, his body jackknifing upward as if pulled by invisible strings, a gasp tearing from his throat that held equal parts ecstasy and terror. The dream—no, the vision—clung to him like a second skin, refusing dismissal with dawn's arrival. Sweat plastered his linen shirt to his broad chest, the fabric translucent and revealing the muscled contours beneath. His heart hammered against his ribs as if trying to escape the confines of mere flesh, having tasted something vaster. Around him, the canvas walls of his spartan tent glowed dully in the single lantern's light, shadows gathering in corners like crouched observers, witnesses to his transformation.

For several heartbeats, Malrik remained frozen in this posture of awakening, his eyes wide and unfocused, seeing not the crude interior of his campaign tent but the lingering afterimages of his dream-self crowned in power. His breathing came in shallow gulps, lungs suddenly inadequate for a man who had, in sleep, commanded the very elements. The camp around him slept on, unaware of the revelation that had descended upon their leader—soldiers' snores and the distant calls of night sentries filtering through canvas as an ordinary counterpoint to the extraordinary transformation occurring within these simple walls.

"Gods," he whispered, though the word emerged as reverence rather than exclamation. His calloused fingers rose to touch his brow where the crown of energy had hovered in his vision—not quite touching his skin but claiming him nonetheless as vessel and instrument. Though no physical trace remained, his flesh remembered the sensation: a pressure both terrible and exquisite, like standing beneath a waterfall of pure potential.

Fragments of the vision replayed themselves behind his eyes—his transformed self standing before masses of kneeling supplicants; his hand extended to reshape mountain ranges with mere intention; his voice carrying across vast distances to be heard by all simultaneously. But more intoxicating than these displays of power had been the feeling that accompanied them—complete certainty, absolute purpose, the end of doubt and limitation that had plagued him since childhood. In that dream-state, he had finally been whole.

The cot creaked beneath him as Malrik swung his feet to the ground, the rough canvas floor cool against his bare soles. The physical sensation anchored him partially in the present, though his mind continued to straddle the boundary between dream and waking. He pushed himself upright, muscles responding with fluid strength, yet now feeling curiously insufficient after experiencing the superhuman capabilities of his mantled dream-self.

Three strides carried him across the tent's narrow confines, then three back, his body seeking movement as his mind raced to integrate what he had experienced. The lantern flame wavered with each pass, sending his shadow leaping against the canvas walls—now elongated to giant proportions, now compressed to mere suggestion, a visual echo of his fluctuating perception of himself.

"She promised this was possible," he murmured, voice pitched below the threshold of those who might be standing guard outside. The *she* remained unnamed even in his private thoughts, a precaution born of years operating in realms where names carried power and listeners might lurk beyond ordinary perception. "A mortal vessel, channeling divine will. Not metaphor but literal transformation."

His heavy boots, discarded beside the cot before sleep claimed him, now called to his restless feet. He pulled them on with methodical movements, the familiar ritual providing structure to thoughts that threatened to spiral beyond his control. The leather was well-worn but meticulously maintained, speaking to a man who understood that greatness required both ambition and attention to practical detail.

"Patience," he reminded himself, though the word chafed against the hunger that the vision had awakened—or rather, intensified, for the seed had been planted long ago. "Preparation. Power without control destroys its vessel."

Malrik's fingers flexed unconsciously, mimicking the gesture that had, in his dream, called lightning from clear skies to strike his enemies. No energy manifested in the tent, yet he felt something respond nonetheless—a subtle current flowing through his veins that hadn't been there before, or perhaps had always existed but only now found recognition in his conscious mind.

The contradiction between his current limitations and promised potential stretched him between frustration and anticipation. The path to his apotheosis remained unclear in specifics, though blindingly certain in destination. He had seen himself transformed, had felt divine energy coursing through prepared flesh, and had experienced the adoration of masses who recognized true power when it manifested before them.

"I was born for this," he said, the words emerging with such conviction that the lantern flame seemed to straighten in response, burning brighter as if acknowledging truth when spoken aloud. "Not farmer, not soldier, not even conqueror—but avatar. Vessel. The bridge between realms that lesser men cannot even perceive. I can fix this world!"

His pacing ceased abruptly as he reached the center of the tent, planting his feet with deliberate firmness as if establishing a foundation for what was to come. The shadows retreated from his sudden stillness, gathering at the edges of the lantern's influence like attendants awaiting command. Malrik pressed his palm against the nearest tent wall, feeling the coarse weave beneath his skin, the tangible barrier between himself and the sleeping world beyond. A world that would wake to find itself fundamentally altered once his transformation was complete.

"I will follow her guidance," he breathed, the words barely disturbing the air yet carrying the weight of absolute commitment. Something shifted in his stance as he spoke—a subtle realignment of shoulders, a slight lift of chin, adjustments that would be imperceptible to casual observation yet fundamentally changed how he occupied space.

The conviction settled into his bones with comfortable finality, as if recognizing its rightful home. Whatever doubts might have troubled him before the vision now seemed trivial concerns belonging to a lesser version of himself—one who had not glimpsed the divine purpose for which he had been shaped through trials and tribulations that had only appeared random.

Outside his tent, the night continued its ordinary progression. Guards changed shifts with quiet efficiency, nocturnal animals foraged at the camp's perimeter, stars wheeled through their ancient patterns overhead—all unaware that the man standing motionless in the canvas shelter had just crossed a threshold from which there could be no return. Not conversion or persuasion, but recognition—the final acceptance of what had always been his destiny, merely awaiting the proper moment to emerge from potential into actuality.

Malrik's eyes, had anyone been present to witness, held a new quality—a focus that looked beyond immediate surroundings to something only he could perceive. His breathing had synchronized with rhythms not quite of this world, each inhalation drawing in more than mere air, each exhalation releasing something of his former, limited self. The lantern light caught the planes of his face, hardening them into angles that suggested something carved rather than born, designed rather than evolved.

The decision, once made, brought unexpected calm. The restless energy that had propelled him from his cot transformed into a steady purpose that radiated from his core. Where minutes before he had paced like a caged predator, now he stood with the stillness of a predator who has sighted prey and entered that perfect moment before the strike—all potential coiled into readiness, all awareness focused to a singular purpose.

"Soon," he promised himself, the word was both an acknowledgment of necessary delay and commitment to inevitable action. Whatever preparation remained before his transformation could be completed, he would fulfill with absolute dedication. Whatever trials stood between his current self and the mantled being he had glimpsed in vision, he would overcome with unwavering resolve.

The single lantern flickered once, then stabilized, its flame burning with unusual steadiness. Shadows that had danced and wavered now held their positions with similar resolve, the entire tent seeming to exhale as decision replaced doubt, as purpose conquered uncertainty.

Malrik, the Gypsy Knight, returned to his simple cot not as the man who had earlier sought sleep, but as the vessel who had accepted his role in a design far grander than mortal comprehension could encompass. His dreams, when sleep reclaimed him, were no longer troubled but filled with anticipation—each one another step along the path that Mirror Mage had so carefully laid before him, leading him willingly toward a transformation that would serve purposes far different from those he imagined in his glorious, manipulated visions.

~ Chapter Thirteen ~
Threadmere

Chapter Thirteen

Threadmere – Night (Auronday, 10 Frostwake, TT 1215)

The Glowturn aurora unfurled across the night sky like a living tapestry. Its ribbons of emerald, violet, and gold stretched from horizon to horizon in undulating waves. Edrin and Lysera climbed the gentle slope of the hillside in reverent silence, their breath forming small clouds in the crisp night air, each step carrying them further from the campfire's mundane light and closer to the celestial display. The hill's crown offered an unobstructed view of the phenomenon—a natural altar from which to witness the sky's sacred communion with earth.

"We should reach higher ground," Lysera had suggested earlier, when the first shimmer of green had appeared above the treeline. "Glowturn nights reveal patterns lost to ordinary vision." Her sight-veils remained wrapped around her wrists, unused—tonight she wished to see with mortal eyes, to share the experience as Edrin would perceive it.

Now they stood side by side at the summit, their shoulders almost touching, faces upturned to the spectacle above. Lysera's features were transformed by the cascading lights—her dusky skin adopting hues of emerald and amethyst as the aurora shifted overhead, her eyes reflecting pinpoints of color that contained entire universes. The wind played through her braids, lifting and settling them in rhythm with the pulsing lights.

"There," she said softly, extending her arm toward a particular configuration where violet light curved and branched. "That formation—it resembles the ancient glyph for revelation. And there—" her finger traced another pattern where gold and green interwove in complex spirals, "—that's reminiscent of the truth-seeking pattern. The old texts speak of Glowturn as the night when the Loom's threads become visible to mortal eyes."

Edrin nodded, though his gaze was now fixed on her face. The lights painted her in otherworldly beauty, but it was the wonder in her expression that held him transfixed—the open vulnerability of someone usually so guarded, the genuine awe softening the lines of tension around her eyes. He had seen her focused in battle, determined in council chambers, steady during crisis, but this unguarded joy was rarer and infinitely more precious.

"The ancients believed Glowturn patterns foretold coming events," she continued, unaware of his diverted attention as she scanned the heavens. "Not as simple prophecy, but as glimpses of the Loom's deeper weavings, usually hidden from view."

"And what do you see in them?" Edrin asked, his voice unexpectedly husky.

Lysera turned then, catching his gaze fixed upon her rather than the celestial display. Something shifted in her expression—surprise yielding to understanding, followed by a gentle warmth that had nothing to do with the aurora's light. "You're missing the show," she said, though her tone held no reproach.

"I'm seeing exactly what matters," he replied simply.

The aurora pulsed overhead, its light intensifying as if responding to the current passing between them. They had stood close before—in battle, in council, in moments of danger and decision—but never with this particular quality of awareness, this acknowledgment of something that had been building since their paths first crossed.

Edrin cleared his throat, suddenly finding the directness of her gaze more challenging than any opponent he had faced. "We should discuss strategy for when we reach the northern settlements," he said, retreating momentarily to safer ground. "If Malrik's pattern holds, he'll target vulnerable communities near ley-line convergences."

"Yes," Lysera agreed, though her eyes suggested she recognized his tactical withdrawal. "Though I suspect his movements may become less predictable now. The mirror-fragments from Misthold suggest evolution in his methods, perhaps he's even abandoning the patterns we have been tracking."

The conversation continued in this vein for several minutes—the comfortable territory of plans and countermeasures, threats and responses. Yet beneath these practical exchanges flowed a deeper current.

Eventually, the discussion faltered, leaving a silence filled only by the whisper of wind and the distant calls of night creatures. The aurora continued its silent dance overhead, now forming a crown of interlaced light directly above the hilltop.

Edrin's hand moved to capture Lysera's, his calloused fingers gentle against her skin. The contact sent warmth up his arm—a completion of circuit, a recognition of pattern.

"Since that day at Red Kiln," he began, his voice low but steady, "when I thought we wouldn't make it out...." The words came with difficulty, not from uncertainty about their truth but from the unfamiliarity of speaking them aloud. His life had been shaped by hammer and flame, by the practical language of survival and craftsmanship—not by the articulation of heart's truth.

Lysera's fingers tightened around his, her touch offering encouragement when words faltered. "I know," she said softly. "I've felt it too. This thread between us, growing stronger with each challenge we face." Her free hand rose to touch his cheek, fingertips tenderly tracing the line of his jaw, conveying what language struggled to contain. "Some patterns reveal themselves slowly, but are no less true for their gradual emergence."

Something shifted in the aurora above them. The separate ribbons of light converged directly overhead, their colors intensified from ethereal translucence to vibrant brilliance. Neither Edrin nor Lysera noticed this change; their awareness narrowed to the diminishing space between them.

"I am not a man of words," Edrin murmured, his free hand moving to cradle the back of her head, fingers threading through her braids. "But I know when something fits perfectly, the way metal yields to proper tempering."

"And I am trained to read patterns," Lysera replied, her lips curving into a smile that contained both amusement and affection. "This one speaks clearly enough."

The distance between them vanished as Edrin leaned forward, his lips finding hers with gentle certainty. The contact sent a current through both of them—not the tentative exploration of a first kiss but the recognition of rightness, of completion long denied but finally acknowledged. Lysera's arms wound around his neck as she responded with equal fervor, the tension of months of unspoken longing dissolving into this single point of connection.

As they kissed, a golden light bloomed between their joined hands—not from the aurora above but from within their very touch. They pulled apart slightly, both looking down in wonder as a glyph materialized in the space between their bodies—a pair of curving lines merging into a spiral of interlocking rings pulsing with warm radiance.

"A love glyph," Lysera whispered, her voice tinged with awe. "I've only read of this in the oldest texts—a spontaneous manifestation when souls recognize their true counterparts."

Edrin stared at the glowing symbol, its light casting his features in gold. "That's a lot of big words. Is this… normal?" he asked, his practical nature reasserting itself even in this extraordinary moment.

Lysera's laughter bubbled up, bright and unexpected. "No, Edrin Ashwood," she replied, her eyes dancing with joy. "This is decidedly not normal. This is rare beyond measure."

The glyph hovered between them for several heartbeats, its light gradually diminishing, sinking into their joined hands, absorbing into their very skin like ink into parchment. When it had fully disappeared, a warm tingle remained where it had touched them—a subtle reminder of what had been made visible, if only briefly.

Edrin looked from their hands to her face, wonder and determination mingling in his expression. "Whatever lies ahead," Edrin continued, his voice gaining strength as the warmth of the vanished glyph settled into his bones, "whatever darkness Malrik or the mirror witch conjure, we face it together." His fingers tightened around Lysera's, the callouses of his blacksmith's hands catching against the smoothness of her skin—different textures united in perfect complement, like the varied threads that together formed the Loom's strength. Above them, the aurora responded to their declaration, its ribbons of light coiling into new configurations, as if acknowledging the bond they had claimed.

"The texts speak of such manifestations as rare beyond counting," Lysera said, her scholar's precision momentarily reasserting itself, though wonder still colored her tone. "Not summoned or created, but revealed as a physical expression of what already exists between souls." She lifted their joined hands, examining the skin where the glyph had disappeared as if expecting to find some visible mark. "The love glyph forms its own channel in the Loom, a current that flows parallel to our individual threads without replacing them."

Edrin considered this with the thoughtful expression he usually reserved for complex forgework. "So we are ourselves, but with… something additional binding us?"

"Precisely," Lysera replied, a smile warming her features at his perfect understanding. "Our separate paths continue, but now intertwined in ways that strengthen both." Her free hand traced an exact copy of the glyph in the air between them, scholar's gestures conveying concepts beyond simple words. "The ancient texts describe it as two streams joining—neither losing its essence, but creating something new in their convergence. It can even be used as a power glyph, to use to power emotional thoughts or even other glyphs."

The aurora shifted again, emerald light dominating as it formed a series of curved lines reminiscent of script written by some celestial hand. Lysera's gaze was drawn upward, her expression sharpening with sudden recognition.

"Edrin, look," she whispered, pointing toward a particular configuration where emerald light formed perfect concentric circles pulsing with rhythmic intensity. "That's the vision glyph—the pattern seers have used for centuries to penetrate beyond surface reality." Her voice quickened with excitement. "And there—" she indicated where violet light formed angular shapes like mountains or blades, "—that's the warrior's mark. I've never seen the aurora form recognizable glyphs."

Edrin followed her gestures, seeing not just beautiful lights but now recognizing patterns within the randomness. "What does it mean?"

"It means we're being shown a path," she replied, her voice hushed with reverence. "Seer and warrior, vision and strength—complementary forces needed to face what comes." Her fingers tightened around his. "Just as we discovered in Misthold, when your protective glyph manifested."

The memory of that moment—his body wounded but his spirit somehow channeling power beyond his understanding—resonated between them. "I still don't understand what happened there," Edrin admitted. "How could I, a smith, possible have created protection strong enough to repel Malrik's phantoms?"

Lysera turned to face him fully, her eyes reflecting starlight and aurora in equal measure. "Perhaps not so simple a blacksmith," she suggested gently. "Your connection to the Loom has always been stronger than you acknowledge. Your family tended the flame glyphs beneath Emberfield for generations—such a legacy leaves its mark."

Their faces were close enough that her breath warmed his skin as she spoke. The intimacy of their position, the shared memory of the kiss that catalyzed the glyph's appearance, created a current between them.

"When did you know?" Edrin asked, his question encompassing more than he could articulate. When had she recognized his potential? When had she begun to feel what now flowed so clearly between them? When had coincidence transformed into destiny?

"Not in a single moment," Lysera replied, understanding his unspoken meaning. "It grew like dawn—imperceptibly at first, then with increasing clarity until I couldn't remember darkness." Her hand rose to trace the line of his jaw, fingers lingering against the stubble that shadowed his skin. "In Threadmere's garden, when we spoke of after…that's when I permitted myself to recognize what I had felt for much longer."

Edrin's arm slipped around her waist, drawing her closer against the night's chill beneath the open sky. "I've never been one for fancy words or grand gestures," he said, the self-deprecation in his tone belied by the simple eloquence of his embrace. "But from that first day at Red Kiln, something in me recognized something in you—like metal recognizing the hammer that will shape it into true form."

A shadow passed across Lysera's expression, brief but unmistakable. "And that's what frightens me," she confessed, her voice barely audible above the wind's gentle murmur. "This path we walk—confronting Malrik, opposing the mirror witch, challenging the manipulation of divine power—it leads to places darker than we can imagine. What if this connection becomes another vulnerability? Another point of leverage against us?"

The question hung between them, giving voice to fears lurking beneath their growing attachment. Edrin's response wasn't immediate. He weighed her words with the seriousness they deserved, refusing easy comfort that might prove hollow.

"Any relationship is a risk," he finally said, his voice carrying the quiet certainty of a man who had survived loss before. "Friendship, loyalty, love—all can be used against us. But they're also our greatest strength." His hand moved to cradle her face, thumb brushing across her cheekbone with gentle precision. "I'd rather face whatever comes with you and Bran beside me, whatever the risk, than walk a safer path alone."

The aurora flared above them, gold and violet ribbons intertwining in a spectacular dance. Lysera's eyes glistened with emotion too complex for simple naming—fear and hope, determination and surrender, all woven together like the light above them.

"Then we walk together," she affirmed, rising on tiptoes to press her lips to his once more.

This kiss held no shock of discovery like the first, no magical manifestation to mark its significance. Instead, it carried something deeper—commitment freely given, choice made with open eyes, a covenant sealed without need for a visible sign. When they parted, something had settled between them, a certainty as tangible as the ground beneath their feet.

The aurora had begun to fade, though ribbons of light still danced across the northern horizon. The night air had grown sharper with cold, reminding them of the world beyond this hillside, of companions waiting below and journeys yet to continue.

"We should return to camp," Lysera said, though reluctance colored her tone. "Bran and Alira will wonder what became of us."

Edrin nodded, his arm remaining around her waist as they turned toward the path downhill. The camp's fire appeared as a distant pinpoint of orange light, small but steady against the darkness. As they descended, stepping carefully over frost-slick stones, neither felt the need to fill the silence with unnecessary words. The glyph might have faded from visibility, but its warmth remained between them—a private current of connection binding them together as they returned to the waiting world below.

#

The campfire at the base of the hill carved a perfect circle of warmth from the night's encroaching chill, its flames dancing with contained enthusiasm that mimicked the more spectacular light show overhead. Bran sat cross-legged on a flat stone, his scales shifting through subtle variations of copper and bronze as they absorbed and reflected the fire's heat. Across from him, Alira leaned forward, her fairykin features caught in the flickering light that emphasized the otherworldly angles of her face while rendering the natural green and gold highlights in her auburn hair as metallic as Bran's scales. Between them lay an assortment of scholarly instruments: Bran's precision-tuned fork resting atop his leather-bound almanac, Alira's collection of small crystal vials arranged by color, and a crude map sketched in ash on the ground before them.

"The pattern suggests three-day intervals," Bran observed, using a slender stick to trace lines connecting various points on the ash map. His voice carried the measured cadence of a lifetime spent in archives. "Each manifestation occurs within eight miles of a known ley-line convergence, but never at the convergence itself—always offset, as if deliberately avoiding the most obvious power nexus."

Alira passed him a flask of spiced wine, its aromatic steam rising to mingle with the woodsmoke. "Deliberate misdirection," she agreed, her fingers briefly brushing his scaled hand in collegial warmth. "Classic fairykin hunting strategy—strike not where prey expects, but where their guard relaxes."

Bran sipped the wine, his vertical pupils contracting slightly at the drink's potency. "These windows," he continued, using his stick to circle several areas on the map, "represent what I call 'convergence windows'—periods when the world grows thin, when boundaries between realms become permeable to certain types of energy." He tapped a specific mark with particular emphasis. "The next significant window opens in four days, when Pyrrhos ascends to his zenith position."

The tuning fork beside him vibrated slightly of its own accord, producing a faint harmonic tone lingering in the air before fading. Bran picked it up, studying the still-vibrating tines with scholarly curiosity. "Interesting. The ley-energy concentrates even as we discuss it."

Alira leaned closer, her healer's pendant—a silver flower with petals pulsing with subtle light—swinging forward from her neck. "If Malrik follows the pattern," she said, pointing to Bran's almanac where certain dates had been marked with red ink, "his next move will come when Pyrrhost—the wandering planet—ascends." Her finger traced the celestial conjunction noted in Bran's precise handwriting. "The traditional time for fire-channeling. IF he is still following a pattern. I am still worried about the decoy at the Red Kiln. Why did someone want us away from Threadmere? What took place while we were gone? Nothing seemed out of place when we returned."

Bran nodded, adjusting his spectacles with a clawed finger. The lenses caught the firelight, momentarily obscuring his purple reptilian eyes behind twin discs of reflected flame. "While I can't be certain, I think he and the mirror mage are still using patterns. The resonance patterns I've been tracking suggest he's attempting to harmonize with divine frequencies—specifically those associated with transformation and fire." He turned a page in his almanac, revealing diagrams of overlapping circles annotated with cramped script. "Pyrrhos and Umbryos—an unstable combination under any circumstances. Together in a single vessel…" He left the thought unfinished, but his scales darkened to a troubled bronze.

"My remedies can counteract physical manifestations," Alira replied, gesturing toward her arranged vials. "Soothing agents for burns, stabilizers for transformation sickness, even temporary ley-line blockers that might disrupt channeling." She selected a crystal vial containing liquid that shifted between silver and black, never fully settling into either color. "But against divine possession? Our options narrow considerably, perhaps to nothingness."

"Possession implies unwillingness," Bran noted, his academic precision reasserting itself. "What we observe in Malrik appears to be willing acceptance—a crucial distinction when considering counteragents."

Alira's lips curved into a sudden smile, unexpected warmth breaking through their serious discussion. "Only you, Bran, could make divine catastrophe sound like a terminology debate among stuffy Archive scholars." She mimicked the stern posture and nasal tone of an elderly archivist: "'The proper classification system for apocalypse is essential to accurate cataloging!'"

Bran's scales rippled with amusement, the closest his lizardfolk physiology came to laughter. "While you healers would simply suggest a tincture of willowbark and bed rest, no doubt." He countered with a surprisingly accurate impression of a fussy court physician: "'This apocalypse presents with severe reality disruption. I recommend increased fluid intake and avoiding strenuous activity for at least three days.'"

Their shared laughter softened the gravity of their discussion. The sound floated upward toward the hill where Edrin and Lysera stood silhouetted against the aurora's light, a brief counterpoint to the weight of responsibility both pairs carried in different ways.

As their amusement faded, Bran returned to the map, his expression growing serious once more. "Protective measures for Edrin and Lysera become increasingly important," he said, tracing a circle around their planned route. "Especially considering

Edrin's unusual manifestation in Misthold. If Malrik perceives him as a direct threat to his ambitions…"

"His recovery progresses well," Alira noted, her healer's instincts surfacing. "The wound itself heals cleanly, but the energy channels…" She hesitated, searching for precise terminology. "Something changed in him during that battle. The protective glyph he generated shouldn't have been possible for someone without formal training. It's as if dormant pathways suddenly activated."

Bran nodded thoughtfully. "I've observed similar anomalies in Lysera's abilities. Her sight extends beyond traditional seer limitations, especially when Edrin is nearby." His clawed finger tapped a rhythm against his almanac as he considered the implications. "Their energies appear to amplify when combined—a resonance effect I haven't documented in any prior research."

They both fell silent as the aurora overhead intensified dramatically, ribbons of gold and violet light converging directly above where their two friends stood on the hilltop. The sudden brilliance cast the entire landscape in ethereal illumination, transforming ordinary trees and rocks into sculptures of light and shadow.

"That's not a standard aurora pattern," Bran observed, his scientific detachment temporarily yielding to wonder as he tilted his head back to study the phenomenon. "The convergence geometry suggests…"

Alira's gaze shifted from the sky to the distant figures on the hill. She smiled. "They're quite close," she said softly, a smile warming her features. "Perhaps discussing battle strategies, though their postures suggest a different kind of negotiation."

Bran followed her gaze, his usual analytical expression softening as he observed the silhouettes merging into a single shape against the brilliant sky. "Interesting timing," he noted, though his tone lacked its usual scientific detachment. "Aurora intensification correlating precisely with their proximity."

"Some things even war can't prevent," Alira murmured, genuine warmth infusing her voice as she watched the distant figures. "Perhaps especially during war, when tomorrow offers no guarantees."

Bran remained silent for a moment, his scales shifting to a contemplative copper hue. When he finally spoke, his voice carried uncharacteristic emotion beneath its measured cadence. "The Archive contains seventeen thousand, eight hundred and forty-two accounts of love discovered during conflict," he said, his perfect recall presenting the number without hesitation. "Statistical analysis suggests such bonds form with particular strength and durability, perhaps because they're forged under circumstances that strip away pretense."

Alira's eyes widened at this unexpectedly poetic observation from her usually pragmatic companion. "That's... remarkably insightful, Bran."

The lizardfolk scholar adjusted his spectacles, seemingly embarrassed by his momentary sentimentality. "Merely empirical observation," he replied, though the warmer bronze of his scales betrayed deeper feeling. "Though I confess the data brings unexpected...satisfaction."

They returned to their maps and notes as the aurora began to fade, their scholarly discussion resuming with renewed purpose. Yet something had shifted subtly in their interaction—a shared recognition of the human concerns that existed alongside their intellectual ones, a reminder that even in darkness, connections formed and strengthened, creating their own kind of light.

#

Deep beneath Mount Cindercore, where the earth's blood ran hot and viscous through veins of stone, Malrik knelt in a chamber carved by neither human hands nor natural erosion. The volcanic cavern breathed around him—exhaling waves of scorching air through cracks in the floor, pulsing with the rhythm of some vast, subterranean heart. Sweat streamed freely down his body, creating rivulets through the ash-paste sigils marking his chest, arms, and face in patterns too dangerous for ordinary language. Gone were the colorful garments and jangling charms of the Gypsy Knight; in their place, ritual garb of unbleached linen hung from his powerful frame, the fabric's edges singed and blackened from proximity to the chamber's unnatural heat.

His dreadlocks had been bound back with a cord woven from volcanic glass fibers, each strand capturing and refracting what little light existed in this buried sanctum. The charm-trinkets that once decorated these locks—mementos of adventures and conquests—had been removed, sacrificed to the preparatory flames hours earlier. Only a single ornament remained: a blackened coin stamped with Pyrrhos' flame sigil, suspended from a lock that fell across his forehead.

With methodical precision that belied the trembling of his overtaxed muscles, Malrik arranged thirteen shards of obsidian in a perfect circle around his kneeling form. Each fragment had been harvested from different volcanic eruptions, each capturing a moment when Pyrrhos' power had manifested in the physical world. The obsidian surfaces reflected Malrik's determined expression in multiple distorted versions—some elongated, others compressed, each revealing a slightly different aspect of the man who had journeyed from adventurer to zealot.

"The circle completes," he murmured, his voice no longer carrying the boisterous humor that had once defined him. Its timbre had changed, deepened, as if the very air he breathed burned away something essentially human with each inhalation. "The paths converge. The vessel awaits."

Rising to his full height, Malrik extended his right arm over the central point of the obsidian circle. With his left hand, he unsheathed a blade of volcanic glass, its edge honed to molecular thinness. Without hesitation, he drew the blade across his palm, opening a precise line through flesh that immediately welled with dark blood.

"Pyrrhos, Flame of Rebellion, Spark of Transformation," he intoned, his voice gaining strength as it filled the chamber. "I offer essence freely given, life that it might fuel greater purpose."

Blood dripped from his outstretched hand, falling toward the heated stone floor. But before it could touch the surface, the droplets hissed and vaporized in mid-air, transforming into suspended crimson mist, defying gravity's insistent pull. The chamber's temperature—already nearly unbearable—climbed higher as the blood-mist began to rotate, forming a slowly spinning vortex above the circle's center.

"I who have walked your paths, I who have carried your mark, I who have seen your truth in dreams and waking visions, I call you forth." Malrik's voice rose, filling the chamber with sound that resonated against the stone walls in harmonics too perfect to be accidental. The ash-paste sigils on his skin began to glow with dull orange light, as if embers had been pressed into his flesh. "Manifest, Pyrrhos. Witness your servant's offering."

The suspended blood-mist ignited suddenly, transforming into a spiral of flame with no fuel yet burned with increasing intensity. The fire coiled upward, expanding and contracting like the breath of some enormous beast. Gradually, it assumed a roughly humanoid shape—a torso of concentrated flame, limbs of licking fire, a head formed from the hottest white-blue core of the conflagration.

The heat radiating from this manifestation should have seared flesh from bone, yet Malrik stood unflinching before it, his eyes reflecting twin columns of fire. The blood from his cut palm continued to feed the apparition, each drop vaporizing and adding to its substance. When the flame-entity reached its full height, towering over Malrik, the chamber fell momentarily silent, the very air consumed by the divine presence.

"I would be your vessel," Malrik declared, his boastful nature transformed into zealous intensity. His arms spread wide in supplication, blood still dripping from his cut palm. "A channel through which your divine will might directly shape this realm. The dreams you sent have prepared me for this purpose, the trials have strengthened my flesh to receive your essence."

The flame-shape of Pyrrhos pulsed, its core brightening. Heat-distortions rippled through the air around it, creating patterns that might have been language written in temperature differentials rather than visible symbols.

Before any response could form, the corners of the chamber darkened impossibly, shadows deepening beyond what the laws of light should permit. These patches of darkness stretched, elongated, flowing across walls and floor like liquid night. As they converged on the wall opposite Pyrrhos' manifestation, they formed the unmistakable outline of a great serpent, its coils extending beyond the chamber's physical boundaries into spaces that existed perpendicular to ordinary reality.

"The flame calls, and shadow answers," came a voice that was not truly a voice—multiple whispers layered upon each other, speaking from different directions simultaneously, some preceding the words, others echoing after, creating a disorienting chorus that originated from no single source. "Always thus, since the first division. Always thus, until the final convergence."

Malrik's expression revealed no surprise at this second manifestation, suggesting he had anticipated—perhaps even planned for—this dual appearance. "Umbryos," he acknowledged, inclining his head toward the shadow-serpent with the same respect he had shown the flame entity. "Keeper of Secrets, Shadow of Transformation. Your presence honors this work."

The shadow rippled along the wall, coils shifting in patterns, suggesting amusement, curiosity, or perhaps hunger. "First rite demands truth," the layered voices intoned, each whisper carrying slightly different emphasis, creating meaning beyond mere words. "No mortal vessel holds two gods for long. Your mind will fracture, your body burn away, consumed from within by opposing forces never meant to coexist within flesh's narrow confines."

The flame-shaped Pyrrhos flared in what might have been agreement or a challenge to Umbryos' assessment. Heat pulsed through the chamber in waves that corresponded to unheard heartbeats.

Malrik's shoulders tightened visibly, the only sign that the shadow god's warning had penetrated his zealous certainty. Yet when he spoke, his voice remained steady, carrying the absolute conviction of one who had moved beyond ordinary fear into the rarified territory of true fanaticism.

"I accept the cost," he stated simply, each word carrying the weight of irrevocable commitment. "Let me wield enough power to stop them before they uncover more. I'll sacrifice what remains of me to preserve the divine order." His blood-smeared hands clenched into fists at his sides, then relaxed, a physical manifestation of his surrender to purpose. "Ashwood and his companions approach understanding that threatens the very foundations of your divine authority. They must be stopped before they learn how to challenge the Loom itself."

The obsidian circle around him began to vibrate, each shard humming at a slightly different frequency. The sound penetrated bone and tissue, resonating with fluids and fibers in ways that threatened the structural integrity of both chamber and flesh.

"Truth accepted," Umbryos' many voices acknowledged, the shadow-serpent's form contracting and expanding with each layered syllable. "Intent recognized."

The flame-entity of Pyrrhos surged forward, its fire-limbs extending toward Malrik in what might have been an embrace or consumption. "Vessel offered," it spoke, its voice the roar of forest fires and the whisper of candle flames combined, "Willingness noted. But readiness…" The fire paused, its core dimming slightly. "Readiness requires further proof."

From the deepest shadow of Umbryos' coiled form, darkness flowed outward, meeting Pyrrhos' extended flames. Where they touched, neither extinguished the other—instead, they combined into something that existed between states, neither fire nor shadow but something that incorporated and transcended both. This amalgamation hung in the air before Malrik's face, rotating slowly like a key seeking its lock.

"Preparation remains," the voices of both entities spoke in rare unison, creating a harmonic that made the chamber's very stones groan under its pressure. "Thresholds uncrossed, sacrifices ungiven. The vessel requires final tempering before it can contain what it seeks to hold."

Malrik stood unmoving in the center of the obsidian circle, arms outstretched, blood still dripping from his cut palm to vaporize before reaching the heated floor. His expression betrayed neither disappointment nor relief at this postponement of his transformation.

"I will complete whatever trials you set," he vowed, the ash-paste sigils on his skin pulsing brighter with each word. "Whatever sacrifice is required, I will offer. Whatever threshold stands between mortality and divine purpose, I will cross."

The flame of Pyrrhos and the shadow of Umbryos maintained their partial manifestations, observing the mortal who offered himself so completely to their purpose. Neither entity revealed what tests they intended, what preparations remained, but in their silent contemplation lay implicit acceptance of Malrik's ultimate suitability, if not his immediate readiness.

And so he waited, perfectly still at the convergence point of fire and shadow, his mortal frame already beginning the transformation that would eventually render him unrecognizable as the man he had once been. His dreadlocks stirred in currents of supernatural energy, the single Pyrrhos charm at his forehead now glowing with internalized fire. His shadow, cast by divine flame upon ancient stone, stretched and contracted in patterns that matched no human form—something shaped less by birth and natural development than by the divine forces that would soon inhabit it completely.

In the perfect stillness of that volcanic chamber, as gods considered their instrument and the instrument offered itself without reservation, Malrik's final fragments of true humanity burned away—sacrificed willingly upon the altar of power he could not truly comprehend, yet craved beyond all reason or self-preservation.

CHAPTER FOURTEEN

Threadmere

Forest

Chapter Fourteen

Threadmere – Night (Commonsday, 1 Emberturn, TT 1215)

Night had descended upon the forest, just outside the Threadmere city ramparts, stars piercing through the canopy in scattered constellations mirroring the glowing embers of their small campfire. Edrin and Lysera sat close enough that their shoulders occasionally brushed, creating tiny currents of warmth between them. The sounds of their companions settling into sleep had faded some time ago, leaving them wrapped in the intimate cocoon of cricket song and crackling flames, their voices naturally falling to whispers.

"That moment in Misthold," Edrin said, poking at the embers with a stick that flared with each touch, "when the phantoms converged, I've never felt such certainty about anything before. It wasn't knowledge or skill that guided my hammer, but something...deeper."

Lysera nodded, her dusky skin absorbing starlight and fire-glow in equal measure. "The Loom works through us in ways we cannot fully comprehend. In that moment, you became a conduit for something ancient and powerful." Her gaze drifted from the fire to his profile, strong and steady against the darkness. "You might have untapped gifts that even the Archive Keepers would struggle to classify."

"I don't like the idea that I can't control something that is within me," Edrin said, more softly toned than Lysera had ever heard him utter.

Their conversation meandered through memories of recent battles, strategies for approaching the northern settlements, and concerns about Malrik's growing power. Yet each exchanged glance held a fraction longer than necessary, each accidental touch lingering just beyond casual contact.

"You almost died there," Lysera said suddenly, her voice softer than before. Her hand moved with deliberate purpose, fingers finding the edge of his collar where it revealed the top of his scar. "May I?"

Edrin nodded, his throat working silently as she gently pushed the fabric aside. The scar carved a jagged path from his collarbone down across his shoulder, disappearing beneath his shirt—a permanent reminder of phantom fire that had nearly claimed his life. Lysera's fingertips traced its raised edge with feather-light precision, her touch carrying both clinical assessment and something far more intimate.

"It's healing well," she murmured, though healing was not truly what occupied her thoughts. The scar tissue felt warm beneath her fingers, thrumming with the quiet pulse of his life-force. "But it will always remind you."

"Some reminders are worth carrying," Edrin replied, his voice rough-edged and low. His own hand rose to capture hers where it rested against his skin, joining her exploration. His calloused palm enveloped her more delicate fingers in the shared discovery of his wound.

The fire popped, sending a spiral of sparks toward the star-strewn sky. In that momentary brightness, their eyes met and held—his dark as rain-soaked earth, hers silver-flecked with insight that came from both oracle gift and woman's knowing. The space between them had grown charged with possibility, with unspoken words, in the slight tightening of his hand around hers.

Edrin's free hand rose to her face, brushing back a braid, his touch lingering against her cheekbone with careful reverence. His thumb traced the delicate arch beneath her eye, mapping a constellation of his own making across her skin.

"Lysera," he said, her name emerging as both question and answer. The single word carried every unvoiced thought that had grown between them since that day in Threadmere's garden—the future they had imagined together, the dream of peace after conflict, the recognition of something rare and precious found amid chaos and danger.

She answered not with words but by leaning forward, erasing the remaining distance between them. Their lips met tentatively at first—a gentle exploration, a question posed in physical form. His mouth was warm and slightly chapped from the wind and winter air. Her lips were softer, yielding yet purposeful as they moved against his. The kiss deepened by unspoken mutual consent, tentative discovery giving way to certainty as her hand moved from his scar to curl around the nape of his neck, fingers threading through his short hair.

Edrin's arms encircled her, drawing her closer with a strength tempered by profound gentleness. The kiss transformed, passion flaring between them like wind through embers. Lysera's body melted against his, years of carefully maintained distance dissolving in the heat of their shared hunger. His hands spanned her back, learning the topography of her body through the layers of her robes, each subtle ridge of her spine a discovery that sent currents of pleasure through his fingertips.

Between their bodies, a soft golden light blossomed—from within their very connection. Their love glyph flickered to momentary life, its pattern more intricate now, spiraling outward from the point where their hearts aligned. It pulsed once, twice, then faded back into invisibility, though its warmth remained, binding them in ways beyond physical touch.

"I've wanted this," Lysera whispered against his mouth, the confession slipping free, "since before I understood what wanting could mean."

Edrin's response came in the form of another kiss, deeper than before, his arms tightening around her as if she might dissolve into mist without his embrace. When they parted, his eyes held a vulnerability she had never seen there—the unflinching warrior, now stripped of armor both physical and emotional.

"Show me," he said simply, the words both request and offering.

Lysera's fingers found the fastenings of his shirt, undoing them with deliberate patience. Each newly revealed inch of skin received her touch, her exploration unhurried despite the urgency building between them. When the garment finally fell away, she paused to study him fully—the broad chest marred by scars both old and new, the strong shoulders that had carried burdens beyond his years, the subtle trembling of contained desire that ran through him like current through wire.

His hands moved with equal care as they found the complex ties of her outer robe, his blunt fingers surprisingly dexterous as they unwound the fabric. She guided him when he hesitated, placing her hands over his to show him how the garment released. Layer by layer, they revealed one another, each disclosure accompanied by murmured appreciation and tender discovery.

When skin finally met skin without barrier, both caught their breath at the sensation—the shock of complete connection, of nothing held back or hidden. Edrin's hands, rough from forge and battle, created exquisite counterpoint against the smoothness of her flesh. Where his touch lingered, trails of golden warmth followed, marking their joining with divine recognition.

"Is this—" he began, his body poised above hers, concern for her comfort momentarily overriding his desire.

"Yes," she interrupted, drawing him down to her. "This is exactly what should be."

They moved together with the synchronicity of those whose souls recognized one another—her body rising to meet his, his strength yielding to her guidance, both discovering the ancient rhythm transcending their individual selves. The glyph flickered strongly between them, its light casting their joined forms in honey-gold illumination extending beyond physical boundaries.

As pleasure built toward its crescendo, Lysera felt something remarkable—thread-like connections forming between them, visible not to ordinary sight but to her seer's perception. These filaments of light wove through and around them, binding their separate energies into a shared pattern. The Loom itself seemed to acknowledge their union, incorporating it into the greater weave that bound all living things.

"I love you," Edrin whispered against her throat, the declaration torn from him in the moment of ultimate vulnerability. "Beyond reason, beyond duty, beyond—"

"I know," she answered, her voice trembling with the same emotion that coursed through her body. "I've always known, even when you didn't. I love you the same."

Their release came as a shared transformation—a simultaneous surrender to joy that sent ripples through the glyph-light surrounding them. For one breathless moment, they existed beyond individual boundaries, their consciousness merged through pleasure into a singular experience transcending flesh. The threads binding them pulled taut, then settled into a permanent pattern.

Afterward, they lay entwined beneath the stars, Edrin's cloak drawn over them against the night's chill. His fingers traced idle patterns on her shoulder as her head rested against his chest, her ear pressed to the steady rhythm of his heart. No further words seemed necessary—their bodies had spoken truths that language could only approximate, had forged connections that would endure beyond whatever trials lay ahead.

The fire burned low, its embers pulsing with hypnotic rhythm. Above them, stars continued their ancient journey across the vault of night, indifferent witnesses to the small miracle, unfolding beneath their distant light—two separate threads in the Loom's vast pattern, now woven into spirals of shared destiny that neither gods nor phantoms could easily unravel.

#

Dawn arrived in gradual degrees, as a slow seeping of light through the forest canopy. The darkness retreated in measured steps, stars fading one by one as the eastern sky lightened from black to deepest blue, then to the pale lavender heralding true morning. Dew gathered on leaf and stem in perfect crystalline beads, transforming ordinary foliage into collections of tiny lenses catching and refracting the strengthening light. Birds began their territorial announcements, their calls overlapping into complex auditory patterns, mapping the forest's invisible boundaries.

Theryn materialized at the clearing's edge as if conjured by the first direct rays of sunlight. His arrival made no sound—the displacement of air around his form too subtle to disturb even the most delicate leaves. He paused at the transition between forest shadow and dawn-lit clearing, his green-gold eyes taking in the scene before him with uncharacteristic softness. The campfire had burned down to white ash and faintly glowing coals, yet its circle of stones still contained memories of last night's warmth. Within this protective ring lay Edrin and Lysera, wrapped in his travel cloak, their bodies entwined in the unmistakable intimacy of new lovers. Even in sleep, Edrin's arm curved protectively around her shoulders, while her head rested in the hollow between his neck and shoulder.

Theryn remained motionless, suspended between his natural inclination toward mischief and a rarer impulse toward respect for this fragile moment. The decision was taken from him by the approach of footsteps—Bran and Alira returning from the nearby stream, their arms laden with provisions for the morning meal. Bran carried a string of small fish wrapped in wet leaves, while

Alira balanced a basket of forest berries and roots she had gathered at first light.

Sensing Theryn's presence, they slowed their approach, gazes following his to the sleeping pair. Bran's scales shifted through subtle variations of copper as understanding dawned, settling into a warm bronze that suggested approval rather than surprise. Beside him, Alira's expression blossomed into a smile of genuine pleasure, intensifying the fairykin's natural luminescence and casting faint rainbow patterns across the dewy grass.

Theryn raised a finger to his lips, the gesture both unnecessary and strangely gentle coming from one who typically delighted in disruption. The three stood in a momentary tableau. Then Theryn pulled something from a pouch near his collar, a multihued rollup of some strange materials that Bran and Alira had never seen, at least not together. He silently lit the rolled-up leaves with a large match and puffed away silently.

"About time," Bran mouthed silently, his reptilian features arranging themselves into what passed for a smile among his kind.

The morning breeze shifted direction, carrying the scent of fresh fish, sweet berries, and a hint of Sythari scentcoil smoke, toward the sleepers. Edrin stirred first, his warrior's instincts registering the presence of others even through the haze of deep slumber. His eyes opened, focusing immediately on Lysera's peaceful face against his shoulder before awareness of their audience registered fully. When recognition came, his body tensed briefly, then deliberately relaxed—a conscious choice to embrace rather than retreat from the vulnerability of their situation.

Lysera woke more gradually, her transition from dreaming to wakefulness flowing like water rather than breaking like glass. She blinked once, twice, her silver-flecked eyes finding Edrin's face before tracking his gaze to their assembled companions. A flush of pink spread across her cheeks, yet she made no move to disentangle herself from his embrace.

"Please," Theryn finally broke the silence, smoke rolling from his mouth around the cigar, his hand flying dramatically to shield his eyes while his fingers remained conspicuously spread to see. "There are innocent creatures present. Think of the squirrels' delicate sensibilities." Despite his teasing words, he maintained a careful position, placing himself between the clearing's open side and the forest path—a sentry disguised as a spectator.

Bran approached the fire circle, setting down his catch and reaching for the water flask with deliberate focus. Yet as he worked, he extracted a small leatherbound almanac from his vest pocket, laying it open beside him with exaggerated casualness.

"Most fortuitous alignment," he remarked to no one in particular, one clawed finger tapping a specific constellation marked on the almanac's star chart. "The Lover's Crown ascends during Heartfire Moon—traditionally associated with unions of unusual auspiciousness." His scales rippled in the lizardfolk equivalent of a private chuckle. "Purely academic observation, of course."

Alira moved with grace, approaching Lysera with a spare cloak draped across her outstretched arms. "Morning brings chill, even after the warmest nights," she said, her voice musical with suppressed delight. She settled the garment around Lysera's shoulders with maternal tenderness, her fingers briefly squeezing the younger woman's arm in silent support.

Edrin sat up, keeping one arm around Lysera while reaching for his discarded shirt with his free hand. His initial embarrassment had transformed into a quiet pride that straightened his spine and lifted his chin. "I wasn't aware we'd called for an audience," he said, though the curve of his mouth softened the words.

"Audience implies advance notification," Theryn quipped, abandoning his fake modesty to collect kindling for rebuilding the fire. "We prefer to think of ourselves as surprised but supportive witnesses to what was, frankly, inevitable." He struck a pose of exaggerated dignity, his cigar sticking out of his upthrust face, wreathed in smoke. "We are professionally discreet. Mostly."

"What our theatrical friend means," Alira interjected, her voice warm with genuine affection, "is that joy found amid darkness deserves celebration, not scrutiny." She extended a handful of berries toward the couple.

Lysera accepted the offering, her initial discomfort melting into grateful recognition of the acceptance radiating from her companions. "Thank you," she said softly.

Bran had already arranged the fish on green-wood sticks propped over the rekindled fire, his efficient movements belying the sideways glances he cast toward Edrin and Lysera. "Statistical analysis suggests relationships formed during periods of heightened danger demonstrate remarkable durability," he observed in his academic tone. "The Archive contains seventeen thousand, eight hundred and forty-two documented cases of—"

"Spare us the research paper, scale-brother," Theryn interrupted, dropping his final load of kindling beside the fire, and dropping the remains of the cigar in the now stoked-up fire. "Some of us haven't had breakfast yet." Despite his words, he raised an imaginary cup in toast toward the couple.

"Seventeen thousand, eight hundred and forty-three now," Bran amended, raising his actual water flask in a mirroring gesture. The liquid caught the morning light, transforming ordinary water into a momentary prism.

As they settled into the comfortable rhythm of morning camp duties, the conversation flowed between practical matters and gentle teasing. Alira shared stories of her first healing expedition, where two scouts had similarly 'discovered mutual appreciation for night watches.' Theryn recounted outlandish tales of romance in unlikely circumstances, each story more improbable than the last, yet delivered with such conviction that laughter rippled through their circle like wind through summer grass.

Throughout it all, Edrin's hand remained connected to Lysera's—sometimes their fingers intertwined, sometimes his palm rested at the small of her back, sometimes her head leaned against his shoulder. These points of contact served as physical manifestations of the threads that had woven between them the previous night, visible as the simple, profound comfort of belonging.

The morning continued to brighten around their small clearing, sunlight finding pathways through leaves in complex and beautiful patterns, mirroring the bonds that held this unlikely family together—forged through shared purpose, mutual protection, and the quiet acknowledgment of the power of their freely chosen connections.

#

The sanctum existed in a perpetual state of twilight, suspended in that liminal space where perception itself became unreliable. Obsidian mirrors lined the circular chamber from floor to vaulted ceiling, their polished surfaces drinking what little light existed rather than reflecting it. Each mirror stood precisely nine handspans apart from its neighbors, creating a perfect mathematical harmony that hummed just below the threshold of hearing. The illusionist moved through this carefully constructed geometry with the fluid precision of one who understood that in such spaces, intention shaped reality as surely as physical action.

Her hands, beautiful despite their unnatural pallor, arranged silver basins in a pattern, mimicking the Great Convergence constellation. Each basin—carved from a single piece of mirror-lotus root, its organic shape preserved but mathematically perfected—contained a different substance. The first three held molten glass dust harvested from lightning strikes on desert sand, its particles suspended in oil pressed from flowers blooming only during lunar eclipses. The middle ring cradled shadow-iron filings, metal mined from veins that had never known sunlight, each microscopic shard capable of storing impressions of darkness itself. The final, innermost vessels contained ember ash collected from the funeral pyres of seven oracles, their remains still carrying echoes of prophetic vision.

"Alignment proceeds according to design," the mirror witch murmured, her voice modulated to harmonize with the chamber's ambient resonance. Her hair—shifting in hues of violet, silver, and blue—moved with subtle autonomy, tendrils occasionally rising against gravity's pull. The fractured planes of her eyes reflected the chamber in countless tiny perspectives, each shard capturing a slightly different version of reality, allowing her to perceive potential outcomes simultaneously.

From a case lined with black velvet, she withdrew a stylus crafted from volcanic glass that had cooled in the blood of a willing sacrifice. The implement felt alive in her grip, its surface temperature fluctuating in response to her touch. With mathematical deliberation, she began inscribing the transference glyph upon the chamber's central floor. Each line connected to multiple others, creating intersections where energy could be redirected, diffused, or intensified according to her will. The glyph's overall structure resembled a spiderweb designed by a mathematician, its precise angles and curves calculated to channel power along predetermined paths.

"The symmetry must be flawless," she whispered to herself, though in this chamber, every utterance became part of the ritual's foundation. Her fractured eyes tracked invisible geometries as she worked, seeing not just the physical markings but the energy patterns they would generate once activated. Where the stylus touched stone, it left behind lines of dark iridescence, somehow inverted—as if the shimmer came not from light's reflection but from its absence.

When the base pattern was complete, the mage stepped carefully to its edge, removing her outer robes to reveal ritual garments beneath—silver-thread embroidery on midnight fabric creating secondary glyphs that would amplify her connection to the powers she sought to manipulate. Around her neck, resting between her breasts, hung the pendant of jagged mirror-shard, its surface constantly shifting between reflection and opacity, never allowing a true image to form.

"Pyrrhos, Flame of Rebellion," she began, her voice never rising above a whisper. "Transformation's catalyst, breaker of stagnant patterns, I call upon your divine fire." Her right hand extended toward the eastern mirror, fingers spreading to resemble a flame's reaching tongues.

Simultaneously, her left hand extended toward the western mirror, though its positioning created a subtly different sigil. "Umbryos, Shadow of Secrets, keeper of hidden truths, revealer of what lies beneath falsehood's veil, I seek your ancient wisdom." The words emerged in the same whispered tone, carrying harmonic undertones that differentiated this invocation from the first, creating two distinct channels of communication despite appearing as a single ritual to casual observation.

The illusion-glyph embedded in her pendant activated, its power manifesting as a thin film of distortion, surrounding her form like heat shimmer above desert sand. This field bent perception itself, ensuring that each divine entity would perceive only their own invocation, believing themselves the sole focus of the sorceress's supplication. The deception required extraordinary control—maintaining two separate energy signatures, two distinct intentions, two independent channels of communication while appearing to direct her full attention to each god separately.

The chamber's temperature fluctuated wildly as the first response came—patches of air near the eastern mirror suddenly heating to nearly unbearable levels while sections near the western mirror grew cold enough to crystallize moisture from the air itself. The ritualist's face revealed nothing but reverent supplication, though behind this mask, calculation and triumph intermingled. They had answered, each unaware of the other's presence, each believing they acted independently.

"Great Pyrrhos," she continued, her right hand now tracing fire-glyphs in the superheated air, "the vessel is prepared as promised—a mortal frame worthy of your divine essence, strengthened through trials of flame, ready to channel your transformative power into the physical realm." Her voice carried perfect conviction, every word concealing her deeper purpose.

To her left, she addressed Umbryos with equally convincing devotion, "Master of Shadows, the worthy vessel awaits your enlightening darkness, his mind expanded through secret knowledge, his spirit aligned with your subtle purposes." Again, no falsehood passed her lips, merely a strategic omission of her complete design.

The obsidian mirrors began to respond, their surfaces rippling like black water disturbed by underwater currents. From the eastern mirrors emanated a deep crimson glow that somehow illuminated the chamber in bloodied hues. The western mirrors exuded tendrils of living shadow moving with deliberate purpose, probing the air as if seeking something specific. Neither manifestation acknowledged the other's presence—the watcher's illusion-field maintaining perfect separation between these opposing divine energies.

With precise timing, she activated the silver basins, each substance beginning to transform according to its nature. The glass dust ignited into suspended points of flame orbiting one another in miniature constellations. The shadow-iron filings liquefied, forming mercurial pools that reflected nothing yet seemed to contain infinite depth. The ember ash rose in spiraling columns, retaining the exact shape of their containers, creating phantom vessels of pure energy.

"The channels are open," the mirror witch intoned, now moving to the center of her inscribed glyph, where all lines converged. "The pathway formed. The offering accepted." Her body became the final component of the complex mechanism—the central conduit through which divine power would flow before being redirected to its intended destination.

Raising both hands overhead, she completed the ritual with a single word that existed in both gods' languages with slightly different meanings: "Converge."

Divine energy surged into the chamber from both sides simultaneously. Pyrrhos' flame manifested as spiraling columns of crimson and gold, carrying the essence of transformation and rebellion. Umbryos' shadow flowed as liquid darkness that held within its currents the secrets of what existed beyond conventional reality. These powers poured into the veiled adept's carefully constructed channels, following the pathways through the mirror-lotus basins and inscribed glyphs.

As the energies passed through her body, her fractured eyes reflected twin flames—one red, one shadow-black—the only outward sign of the immense power temporarily flowing through her form. With exquisite control, she directed the majority of this divine essence toward its intended purpose, accumulating it in the central convergence point of her glyph to be transported into Malrik's waiting form.

Yet at the critical moment of transference, her fingers traced one final, nearly imperceptible sigil that modified the flow. This subtle alteration created a secondary channel—a hairline fracture in the conduit, diverting a small but significant portion of the combined divine energies. This siphoned power flowed into the jagged mirror-shard pendant at her throat, where it was stored rather than transferred, accumulating in crystalline lattices invisible to any perception save her own fractured vision.

To the watching gods, the ritual appeared to function exactly as intended—divine power flowing through a willing priestess to empower their chosen mortal vessel. Neither would detect the subtle diminishment of their contribution, the careful skimming of power for the mirror-worker's personal use. Her face remained a mask of reverent concentration, revealing nothing of the triumphant knowledge that with each such ritual, her own reservoir of stolen divinity grew incrementally larger.

The transfer completed with a pulse of gravity-defying energy, causing the practitioner's hair and robes to float around her before settling as normal physical laws reasserted themselves. The mirrors gradually stilled, their surfaces returning to ordinary obsidian reflection as divine presence withdrew, each god satisfied with their individual contribution to a working neither realized was shared.

The mirror mage stood motionless at the center of her complex glyph, allowing her breathing to stabilize as the pendant against her chest pulsed with newly acquired power. Her lips curved into a smile of perfect satisfaction.

#

The valley floor lay stripped of life, a concave expanse of pale dust and eroded stone resembling some massive, long-extinct creature's eye. No plant had taken root here in centuries, no animal willingly crossed its barren expanse. Even insects avoided the subtle wrongness that permeated the soil—an ancient geological memory of violence too profound for organic life to forget. At the valley's center, precisely positioned according to calculations that accounted for ley-line convergences and stellar alignments, Malrik knelt within a perfect circle of volcanic ash. His ceremonial garb—simple linen stained with symbolic patterns rendered in his own blood—hung loose on his frame, as if his body had already begun surrendering unnecessary mass in preparation for what was to come.

His dreadlocks had been bound with copper wire, each metal coil inscribed with microscopic glyphs that would serve as conduits for the coming transformation. The Pyrrhos charm at his forehead—a blackened coin stamped with the flame sigil—pulsed with dull heat against his skin. Malrik's breathing maintained a steady, controlled rhythm as he waited—arms outstretched in supplication, eyes fixed on some middle distance beyond ordinary perception.

The power arrived without warning—not gradually but as a sudden cataclysm. One moment, Malrik knelt in anticipation; the next, his body convulsed violently, causing his spine to arch beyond natural limits, bones creaking with stress that would have shattered an ordinary human skeletal structure. A sound tore from his throat—not quite scream, not quite laughter—as divine energy surged through channels never meant to accommodate such magnitude of power. His muscles spasmed in rapid sequence, skin rippling as if seeking escape.

"Yes," he managed through clenched teeth, blood seeping from gums where his bite had broken flesh. "Come through me. Remake me. Use me." Each word emerged strained and distorted, his vocal cords already transforming.

The first visible change came as hairline fractures across his exposed skin—delicate patterns like frozen lightning spreading from his sternum up his chest, his neck, and down his arms. Within seconds, these fissures deepened, widened, revealing something beneath. Not blood as humans know it, but a substance that resembled molten metal—thick orange-red fluid pulsing with its own rhythmic life, catching sunlight and intensifying it until looking directly at Malrik became painful to mortal eyes.

As the transformation accelerated, distinct patterns emerged in how the divine energies manifested. From his waist upward, Pyrrhos' influence dominated—fire, literal and metaphorical, racing across his skin, dancing between his fingers, occasionally erupting from his mouth when he exhaled. His upper body gleamed with sweat that sizzled into steam upon contact with air. The muscles beneath his cracked skin swelled with unnatural density, fibers restructuring themselves into something that could withstand and channel the heat that would have reduced ordinary flesh to ash.

Below his waist, Umbryos' shadow held sway. Darkness clung to his legs not as an absence of light but as a presence of something consuming illumination. These shadows moved with deliberate, sentient purpose, occasionally extending beyond his physical form. Where they touched the ground, frost crystallized despite the day's warmth—tiny geometric patterns sublimating directly from solid to vapor without passing through the liquid state.

The valley itself responded to this transformation—dust rising in spiral patterns around the ash circle, stones vibrating with increasing intensity until they cracked from internal pressure. The ground beneath Malrik began to change, ordinary dirt and rock transforming into obsidian as the extreme heat from his upper body melted silicates while the unnatural cold from below rapidly cooled the resulting liquid into glass. The transformation spread outward in concentric rings, creating a platform of black mirror reflecting the sky above in fractured, distorted images.

Malrik's screams had transformed completely now—what emerged from his throat contained both the roar of wildfire and the whisper of darkest caverns. His face, once handsome in its rugged way, had become a landscape of cracks, the features still recognizable but rendered inhuman by the power restructuring them from within. His eyes, previously brown, now burned with orange flame that occasionally leaked from the corners like molten tears.

"I am...becoming," he declared, his voice resonating at frequencies that caused loose stones to skitter across the newly formed obsidian platform. Each word emerged weighted with unnatural authority—not merely the statement of a man but the declaration of something that existed partially beyond physical limitations. "But... incomplete."

He felt it—the frustrating boundary beyond which his transformation would not progress. Something held back the final convergence of divine energies that would have rendered him a true avatar of both gods. Instead, he balanced at the threshold between mortal and divine, powerful beyond human measure yet still constrained within a partially human form. Unknown to him, this limitation was the mirror illusionist's design—a carefully calculated restriction keeping him dependent upon her for further advancement while preserving her own control over his developing powers.

Malrik rose unsteadily to his feet, testing this new body that both was and was not entirely his own. His movements contained curious duality—his upper body fluid and quick like dancing flame, his lower half deliberate and weighty. The linen garment that had covered him fell away in charred tatters, revealing his transformed flesh in full—a living anatomy lesson in divine possession. His chest heaved with breaths that drew in more than air, inhaling ambient energy from the surroundings themselves.

Extending one arm, Malrik focused his attention on a boulder at the valley's edge, some thirty paces distant. His fingers spread, trembling with effort as he channeled his new abilities for the first time. Fire spiraled down his arm, intertwining with shadow at his fingertips, the opposing forces creating something that existed between states—flame-edged darkness that leapt from his hand toward the designated target. The boulder didn't merely shatter but transformed—half melting into slag, half dissolving into shadow-substance that evaporated upward like negative steam.

A laugh bubbled from Malrik's fractured lips, genuine joy mingling with darker satisfaction. He turned, directing his attention to a different target, then another, each demonstration of power more controlled than the last as he learned the parameters of his abilities. The valley floor became his training ground, its features reshaped according to his experimental whims. Here, a pillar of obsidian flash-formed from sand, there a crater where matter had been partially erased from reality.

As his confidence grew, his thoughts crystallized around the purpose that had driven him to accept this transformation. His fragmented consciousness—part Malrik, part divine influence—coalesced around a single directive that burned clearer than any other consideration.

"Edrin Ashwood," he spoke, the name emerging as both curse and promise. "Loom-questioner. Pattern-breaker." Each designation carried specific contempt, divine concern transmuted through his mortal understanding into personal vendetta. "You seek knowledge not meant for mortal minds. You push against boundaries established since the first weaving." His hand clenched into a fist, molten blood seeping from fresh cracks in his knuckles to sizzle against the obsidian beneath his feet. "Your questions threaten the very foundations of divine authority."

The twin powers within him responded to this focused intention, shadow and flame momentarily aligning in perfect harmony rather than opposition. Malrik's form stabilized, the constant pain of his partial transformation receding as purpose overrode physical discomfort. He turned his gaze northward, somehow sensing the direction where Edrin and his companions traveled, his fractured perception extending beyond normal limitations to glimpse threads of possibility, paths of pursuit.

"The gods have spoken," he declared to the empty valley, his voice carrying the double resonance of divine influence. "The pattern must be preserved. The questions must cease. The Loom's workings remain the province of its creators alone." He began walking, each footstep burning through the twigs and grasses below his feet, melting the very rock below it and immediately cooling, leaving behind perfectly formed obsidian prints in the dry soil. "I will be the instrument of their judgment. I will be the silence that answers your questions, Ashwood."

As he reached the valley's edge, Malrik cast one final glance at the transformed landscape behind him. Then he turned away, his pace increasing beyond human capability, fire and shadow alternately propelling him forward as he began his hunt. With each stride, his new purpose crystallized further: find Edrin Ashwood, stop his dangerous inquiries by whatever means necessary, preserve, and perhaps finalize the divine influences that had elevated Malrik to his current state.

Behind him, the obsidian platform at the valley's center cracked with a sound like distant thunder, a perfect fracture line splitting the mirror-like surface precisely in half—fire-touched stone on one side, shadow-cooled glass on the other. The division spread outward, creating a boundary mapping the unstable duality now housed within Malrik's transformed flesh—a physical warning of the inherent instability in what he had become, visible to anyone who might follow but invisible to the vessel himself.

CHAPTER FIFTEEN

Emberfield

Chapter Fifteen

Emberfield – Dawn (Auronday, 28 Newthread, TT 1215)

The wedding planning was almost at an end as dawn crept across Emberfield in delicate increments, frost transmuting into diamond dust wherever morning light touched the winter landscape. The partially rebuilt village awakened not to another day of vigilance against unseen threats, but to the hum through frost-stiffened air like the resonance of a perfectly struck bell. Today marked Hearthbless, the day before The Deathless Toast—the winter festival of community and hearth-sharing. Still, more significantly, it would witness the union of Edrin Ashwood and Lysera Vayesh, a commitment forged through trials that had tempered their bond rather than fracturing it. Throughout the village, smoke rose from chimneys in straight, determined columns, each household contributing to the thin veil that hung above Emberfield like protective gauze against the crystal-clear sky.

In the village smithy, Edrin worked with methodical focus, hammering heated iron with rhythms that spoke of years spent mastering his craft. Each blow shaped simple ornaments meant to encircle the ceremonial hearth—interlinked rings and spirals symbolizing continuity through winter's harshest months. His movements were measured, controlled, yet beneath this studied precision lay a current of anticipation that occasionally lifted his gaze toward the door, as if expecting Lysera to materialize within the frame. When he wiped sweat from his brow despite the winter chill, his sleeve fell back to reveal the faint golden outline of the love glyph—now a permanent marking upon his skin that pulsed gently with his heartbeat.

"These will cool in time," he murmured to himself, setting aside another completed hearth-ring. He tested its weight in his palm—substantial enough to endure, light enough to hang from the ceremonial garlands without bending the boughs. Like their bond, he thought but did not say aloud, strong yet never burdensome.

Across the village in the communal hall, Lysera's fingers traced each ritual glyph with seer's precision. Her touch lingered on ancient symbols that had guided communities through winter darkness since before written history. Silver-flecked eyes narrowed in concentration as she verified each marking, occasionally adjusting a line or deepening a curve where time had softened the original intent. The glyphs hummed beneath her fingers—recognition passing between known entities—as if acknowledging her dual role as both oracle and bride. Her burnished skin absorbed the hall's shadows while reflecting pinpoints of light from the hearth fire, creating an effect of stars contained within human form.

"The protection sequence needs strengthening," she noted to an elder who worked nearby, hanging frost-tipped evergreen boughs in symmetrical arcs. "The binding glyphs are clear, but these warning marks have faded." Her voice carried no anxiety, only practical assessment—the habit of one who had faced genuine danger and learned to address it directly rather than through fear.

The elder nodded respectfully, having witnessed the oracle's abilities during the phantom attack months earlier. "We can deepen them with fresh pigment before the ceremony. The red ochre was prepared yesterday."

Lysera acknowledged this with a slight smile, her attention already moving to the hearth itself, where she arranged winter garlands in patterns that complemented the glyph sequences. Each placement was deliberate, transforming mere decoration into a continuation of the protective symbols encircling the communal space. Though her hands remained steady with ritual purpose, occasionally her gaze lifted toward the smithy's direction, a momentary softening crossing her features before returning to focus.

Near the hall's entrance, Bran hunched over a table strewn with parchments, his scaled fingers marking precise notations on a schedule of ceremonial participants. His spectacles reflected the morning light in twin discs that obscured his reptilian eyes, giving him the appearance of some ancient scholar risen from dusty archives rather than living flesh. His tail occasionally twitched with satisfaction when a particularly complex sequence aligned properly.

"Elder Mareneth will lead the hearth-lighting," he muttered to himself, scales shifting to a thoughtful bronze. "Followed by the smith's guild offering, then the oracle's blessing." His clawed finger tapped the parchment precisely. "Seventeen minutes for traditional recitations, allowing for Elder Fareneth's tendency to extend the winter blessing by approximately thirty-eight seconds."

Beyond his meticulous scheduling lay a deeper purpose—each arrangement calculated to maximize protective influences, each transition timed to align with subtle energy patterns strengthening the community bonds formed during Hearthbless. Though he would never frame his work in such sentimental terms, beneath Bran's scientific precision lay genuine care for the human rituals that bound people together through winter's darkest days.

In the herb garden behind the hall—protected from frost by cleverly arranged windbreaks—Alira knelt among dormant plants, her fairykin senses detecting vital energy still pulsing beneath winter-hardened stems. Her fingers, graceful despite their strength, harvested sprigs of evergreen maryvine and winter yesch, each cut made with murmured thanks to the plants for offering their essence. Her auburn hair caught morning light in highlights of green and gold, occasionally humming melodies that seemed to make the frozen soil itself respond with subtle warming.

"Healing and protection," she whispered, separating herbs into careful bundles. "Remembrance and cleansing." Each grouping would be woven into ceremonial ribbons—not merely as decoration but as practical enchantment, infusing the celebration with properties extending beyond symbolic meaning.

When she spotted a sprig of winter bloom—rare and unexpected in this season—her smile deepened with recognition of the omen. "Hope flowering in unlikely places," she said to herself, carefully adding it to a special bundle meant for the bride's ribbons.

As morning advanced, the four companions converged around a rough-hewn table in the hall's side chamber, steam rising from bowls of porridge sweetened with preserved berries and drizzled with precious honey reserved for significant occasions. Their circle formed naturally, each finding their place without discussion—a family created through choice rather than birth.

"The iron work is nearly complete," Edrin reported between careful spoonfuls, his blacksmith's hands incongruously gentle around the carved wooden bowl. "Six more hearth-rings, then the ceremonial hammer for the striking."

"The elders' procession is arranged," Bran contributed, his scales shifting to pleased copper. "Exactly seventy-three minutes for the complete ritual, accounting for seasonal adjustments and the additional wedding elements." He adjusted his spectacles with scholarly precision. "Statistically, winter weddings during Hearthbless show sixteen percent greater longevity than those conducted in other seasons."

Alira's laugh rippled through the serious conversation like sunlight through ice. "Always the romantic, Bran." She passed a small pot of honey toward Lysera with subtle encouragement. "The ribbons will be ready by midday. I found winter bloom for your hair wreath—quite remarkable for this season."

Lysera accepted the honey with a grateful nod, her fingers briefly touching Alira's in quiet acknowledgment. "The hall glyphs need strengthening, but the foundation patterns remain strong." Her free hand found Edrin's beneath the table, their fingers intertwining in habitual connection. "The warning markers are faded, however."

A shadow passed across Edrin's features, warrior's instinct momentarily overtaking the bridegroom's anticipation. "How faded?"

"Not through neglect," Lysera assured him, her thumb tracing circles against his palm. "Just time's natural erosion. They'll be renewed before the ceremony."

Beyond their quiet conversation, the communal hall buzzed with increasing activity. Villagers arrived bearing covered dishes for the feast—steaming pots of winter root stew, loaves of dark bread studded with preserved fruits, and precious wheels of aged cheese brought out from storage for this special occasion. Children darted between adults, their high voices practicing traditional songs, occasionally dissolving into giggles when lyrics were forgotten or notes went astray. The metallic scent of ribbon dye mingled with the rich aromas of food and the sharp, clean smell of frost-covered boughs being brought indoors to decorate the space.

As they finished their meal, Bran disappeared briefly, returning with a carefully folded bundle of cloth. With uncharacteristic ceremony, he placed it before Edrin, scales shifting to deep bronze, signaling his genuine emotion.

"I completed this last night," he explained, gesturing for Edrin to unfold the fabric. "Statistical analysis of your mutual attraction manifestations suggested a pattern."

Edrin carefully opened the bundle to reveal a mantle of deep charcoal gray, its borders embroidered with silver thread that caught the light in subtle patterns. But the true significance lay at its center, where Edrin and Lysera's two lines, becoming intertwined in spirals, had been woven into the fabric with threads that contained captured sunlight, glowing faintly with inner radiance.

"The glyph manifested naturally between you," Bran continued, his voice taking on the cadence of a lecture. "But harnessing its power deliberately requires practice. This mantle should facilitate channeling—creating resonance between the woven symbol and the mark upon your skin."

Edrin touched the fabric with blacksmith's fingers, suddenly hesitant, as if handling something far more precious than the metals he formed daily. "This is…remarkable craftsmanship."

"The Archive texts suggest such focuses enhance control by approximately sixty-three percent," Bran added, scales rippling with pleased embarrassment at Edrin's evident appreciation. "Given recent encounters, improving your ability with glyph magic seems prudent, alongside more conventional combat training."

"He means you'll need both hammer and magic and the learning of them to keep us all breathing," Theryn's voice interjected from the doorway, his lean form materializing as if conjured by the conversation. His usual sardonic smile carried genuine affection as he approached. "Though if you manage to marry our lovely seer without disaster striking, I'll consider it victory enough for one day."

Lysera rose gracefully, her fingers trailing across Edrin's shoulder as she moved toward the main hall. "Then we should continue preparations to ensure that victory." The silver in her eyes caught morning light as she glanced back at Edrin, her expression conveying everything words could not—determination and hope, commitment and desire, all woven together as intricately as the glyph that connected them.

As they dispersed to their tasks, each carried awareness that today's celebration existed as fragile light against gathering darkness. Yet in that very fragility lay its precious nature—a moment claimed for joy amid uncertainty, a declaration that some bonds would endure whatever trials approached from beyond Emberfield's frost-covered boundaries.

#

Midday sun transformed Emberfield into a tableau of brilliant contrasts—sharp shadows cast by buildings across pristine snow, steam rising from freshly baked bread cooling on windowsills, evergreen boughs drooping beneath the weight of melting frost that dripped like measured time. The communal hall had become the village's beating heart, its doors constantly opening to admit villagers bearing additional contributions to the feast or children racing in with armfuls of decorative pine cones and winter berries. Within this flurry of preparation and anticipation, few noticed the lean figure approaching from the northern perimeter, his usual fluid grace replaced by the measured precision of a man carrying unwelcome knowledge.

Theryn moved against the current of celebratory activity, his eyes scanning methodically rather than dancing with their usual mischief. The snow retained no trace of his passing—a skill earned through years among the nomadic Stone Caravan, where being tracked often meant being hunted. He located Edrin near the smithy, carrying the last of the ceremonial hearth-rings toward the communal hall, the mantle Bran had crafted folded carefully over one arm.

"A word," Theryn said, his voice pitched for Edrin's ears alone. The playful cadence of his speech had flattened into something harder, more utilitarian. He was chewing on the end of another one of his smoking rollups. His hand closed briefly around Edrin's forearm, fingers pressing with deliberate urgency. "Not here."

Edrin recognized the shift immediately. The transformation from companion to scout that occurred whenever Theryn returned from the boundaries with something worth reporting. Without question, he altered his trajectory, nodding toward an elderly woman, who questioned the placement of his forged pieces.

"The master smith can advise," he told her, already moving toward the narrow space between storehouses where their conversation would remain private.

In this shadowed corridor, shielded from both sunlight and curious eyes, Theryn's posture changed completely. The performer's flourishes vanished, replaced by the economical movements of someone conserving energy for potential action. His green-gold eyes held Edrin's with unusual directness. He took the time to strike one of his large matches, produced from one of the many pouches around his waist, lighting the rollup. Edrin watched him, realizing that there wasn't something happening right at that moment.

"Someone's watching us," he stated without preamble. "Not travelers passing through, not hunters tracking game. Watchers.

Counters." His finger tapped against his thigh in unconscious rhythm as he spoke, "Three observation points were established along the eastern approach. Another at the north crossroads."

"How long?" Edrin asked, his voice lowering to match Theryn's tone. The ceremonial hearth-rings in his hands—symbols of community and celebration—suddenly felt incongruously festive against the serious exchange.

"Recent. Within a day. Snow hasn't completely settled into the depressions." Theryn's gaze flicked toward the main thoroughfare, calculating distances and sightlines with instinctive precision. "Found broken twigs arranged in patterns too deliberate for animal passage. Displaced snow where someone knelt for extended observation. And the silence." He paused, eyes narrowing at the memory. "The eastern thicket should be loud with winter birds. Instead, absolute stillness, as if something had established territory that even animals recognize as dangerous."

Edrin's jaw tightened, muscles flexing beneath paled skin despite the cold that should have flushed it. "I'll be flamed if I am going to let all of this rebuilding get destroyed again. Get the others," he said after a moment's consideration. "Back room of the hall. Quietly."

As Theryn slipped away with characteristic soundlessness, smoke trailing from the cigar, Edrin completed his delivery of the hearth-rings to the communal hall, his external demeanor betraying nothing of the concern now coiling in his chest. He moved through the celebratory preparations with practiced smiles and brief acknowledgments, his awareness simultaneously tracking the positions of his companions throughout the village. Lysera by the hearth, fingers tracing freshly renewed glyphs. Bran consulted with village elders near the main table. Alira weaved ribbons with a group of young women, her fairykin features alight with shared laughter.

When they finally converged in the back room—ostensibly to discuss final wedding arrangements—the transformation was immediate. Laughter lines smoothed into vigilant assessment; festive anticipation yielded to strategic calculation. Bran spread maps across the rough-hewn table while Theryn weighted the corners with iron weights borrowed from the stores. The parchment bore Bran's meticulous notations of ley-line convergences, defensive positions, and terrain features around Emberfield, each marking precise to the point of obsession.

"Show me," Bran instructed without preamble, his scales shifting to the deep copper that signaled intense concentration. His clawed finger hovered above the eastern section of the map, waiting for Theryn's information.

Theryn leaned forward, marking locations with small pieces of charcoal. "Here, where the creek bends before the pine stand. Signs of extended observation—broken ice where someone collected water, disturbed snow beneath the largest pine. And here—" his finger moved to another position, "—someone spent hours watching the main approach. Found a carved marking in the bark of an elm—not a glyph, but deliberate. Three vertical lines intersected by two horizontal."

Bran's eyes narrowed behind his spectacles, their vertical pupils contracting as he compared this information against internal records. "Similar markings were found after the Misthold incident," he noted, scales darkening with concern. "And again at the northern settlements before the ambush at Widow's Fork."

"Malrik's scouts," Lysera concluded, her voice steady despite the implications. Her fingers unconsciously found the protective glyph she wore at her wrist—a silver band inscribed with symbols specific to foresight and warning. "Marking potential approaches, cataloging defenses."

Alira stepped closer, her usual gentle demeanor overlaid with the focused intensity of a healer assessing potential wounds before they occurred. "The outer boundaries need reinforcement," she stated, fairykin practicality asserting itself. "I can prepare warning wards—not strong enough to prevent entry, but sufficient to alert us to incursion."

"How many?" Edrin asked Theryn, his hand resting beside his hammer on the table—a blacksmith's tool transformed by context into a warrior's weapon. The contrast between his formal wedding attire—his new hooded mantle, and the hard-set determination in his expression—created dissonance embodying their current circumstance: joy and danger occupying the same moment, neither negating the other.

Theryn's mouth tightened around the rollup. "Can't be certain. Three, maybe four distinct movement patterns. They travel light, avoid breaking branches or leaving obvious footprints. Semi-professional." He traced several potential approach routes on the map. "Their discipline suggests training beyond common bandits or opportunists."

"Divinely sanctioned," Bran murmured, almost to himself. His claws tapped a rhythm against the parchment as he calculated possibilities. "The timing cannot be coincidental. Hearthbless represents significant community bonding. A power nexus that would attract attention from those attuned to such patterns."

"And our wedding must represent a threat to Malrik's master, or masters," Edrin concluded, the statement neither a question nor a boast but a simple acknowledgment of fact. The love glyph visible at his wrist pulsed once, responding to the protective intensity as he studied the map.

Lysera moved to stand beside him, her shoulder touching his in quiet solidarity. "The binding of our energies creates patterns they can't fully control or predict," she confirmed, silver flecks in her eyes catching the room's dim light. "That uncertainty threatens their perception of divine authority."

Alira began sorting through the small pouches she carried habitually at her belt—healer's instinct preparing for trouble before it arrived. "I need winterbane and frost-thistle for effective boundary wards," she said, mentally inventorying available resources. "And silver dust for the sigil anchors."

"I can provide the metal components," Edrin offered, already calculating how quickly he could adapt ceremonial pieces to defensive purposes if necessary.

Their voices lowered further as they developed preliminary strategies, aware of the celebration continuing beyond the room's closed door. Children's laughter occasionally filtered through weathered wood, the sound creating a poignant counterpoint to their grave discussion of defensive positions and evacuation routes. Outside, the midday sun began its gradual descent. Shadows lengthened across Emberfield's snow-covered paths, stretching toward the eastern thicket where something waited, watching, counting, planning.

"We should inform the village elders," Bran suggested, his academic detachment framing a practical response to a potential crisis. "Quietly, without causing general alarm."

"I'll speak with the head elder," Edrin agreed, straightening with a decision made. "The rest of you continue preparations, both for the ceremony and for whatever approaches from the east. We'll celebrate today and face tomorrow when it arrives." His hand found Lysera's, their fingers interlocking with the practiced ease of those whose connection had been tested repeatedly by adversity.

"And if they come before tomorrow?" Theryn asked, his question carrying no fear, merely tactical consideration.

Edrin's gaze fell to the mantle, the woven love glyph catching light even in the room's dimness. "Then they'll discover we're as dangerous celebrating as we are fighting," he answered, the quiet certainty in his voice lending weight to words that might otherwise have seemed mere bravado.

#

The late afternoon light angled through the communal hall's windows in amber shafts, transforming ordinary dust motes into suspended galaxies. The space had completed its metamorphosis from repaired and rebuilt village gathering place to ceremonial sanctum—evergreen garlands hung in precise arcs that echoed the protective glyphs renewed along the walls, iron hearth-rings interspersed among winter blooms created patterns of symbolic significance beyond mere decoration, and ribbons in blue and silver caught the light in rhythmic flashes as they stirred in the gentle heat rising from the central fire. Villagers had taken their places around the hall's perimeter, arranged according to Bran's meticulously calculated procession order, while village elders stood near the hearth where the binding would take place. Edrin and Lysera approached from opposite sides of the circle, practicing the steps that would bring them together before the ceremonial flame.

Despite the lingering tension from Theryn's discoveries, Edrin moved with surprising grace for a man more accustomed to hammer and anvil than ceremonial proceedings. His formal attire—indigo tunic embroidered with subtle glyphs of union and protection—had been a gift from the village seamstress, its cut emphasizing the strength in his shoulders while softening the warrior's vigilance that had become his habitual posture. The mantle Bran made remained folded at the circle's edge, waiting for the moment in the ceremony when it would be draped across both their shoulders to symbolize shared path and purpose.

Across the circle, Lysera advanced with measured steps, her dusky skin absorbing light while her silver-embroidered robe reflected it, creating an effect of stars moving through darkness. The ceremonial veils—not the sight-veils of her oracular practice but fine-spun silver meant to represent the transition from solitary sight to shared vision—floated around her face like morning mist, occasionally revealing glimpses of her serene expression. Winter bloom had been woven into her braids, the rare white flowers forming a crown, and captured and intensified by the afternoon sunlight.

"When you reach the center stones," intoned the head elder, a white-haired woman whose authority rested in both age and wisdom, "you'll stop at arm's length. Edrin will present the forged binding ring, Lysera will respond with the woven truth-cord."

They followed the instructions with appropriate solemnity, their genuine emotion flowing beneath their practiced movements. When their fingers brushed during the symbolic exchange, the love glyph at Edrin's wrist pulsed once, visible even through his sleeve as a momentary golden light. Lysera's answering smile carried warmth that transcended the ceremony's formality—a genuine connection existing alongside and within the ritual framework.

"Then you'll speak the traditional words of claiming," continued the elder. "Edrin first, then—"

The change came with the suddenness of lightning striking from a clear sky. One moment, Lysera stood poised and present; the next, her body went rigid, her eyes widening as silver luminescence flooded her irises, obliterating pupil and white in uniform metallic sheen. The ceremonial veils around her face billowed outward, their delicate fabric suddenly resembling smoke more than cloth.

"Lysera?" Edrin's voice cut through the startled murmurs, his hand reaching for hers as he recognized the signs of vision-onset. Before his fingers could make contact, her body began to tremble, fine vibrations escalating into visible shaking that threatened her balance.

"Fire," she whispered, the word emerging with eerie resonance that belonged to no human vocal range. "Fire-consuming offerings meant for hearth gods. Fire devouring celebration. Fire that speaks with a human tongue."

Around the hall, rehearsal participants drew back instinctively, creating a widening circle around the seer whose perception had suddenly broken free from present reality. Only Edrin moved toward rather than away, his protective instinct overriding caution. Bran and Alira converged from different directions, their expressions revealing recognition of vision-state but concern at its unexpected violence.

Lysera's arms extended outward, fingers splayed as if pushing against invisible barriers. Her head tilted backward at an angle that suggested her gaze penetrated the ceiling, seeing something beyond the physical limitations of wood and thatch. When she spoke again, her voice carried layers—her own intertwined with something older, deeper, resonating in listeners' bones rather than merely their ears.

"Roaring destruction consuming decorated timber. Support beams blackening, weakening, surrendering to gravity's demand. Children's voices transformed from song to terror." Each statement emerged as a separate revelation, fragments of future possibility delivered without context or chronology. "Shadowed figures moving through smoke like fish through water, their purpose singular and terrible. A blade of obsidian seeking specific blood. A glyph activated by sacrifice rather than celebration."

Edrin reached her as her knees began to buckle, his arms encircling her, preventing her from collapsing. Against his chest, her body continued its violent trembling, eyes fixed on horrors visible only to her oracular sight. Her fingers clutched his tunic with desperate strength, anchoring herself against the vision that threatened to sweep her consciousness away completely.

"Emberfield burning. Wedding feast becoming funeral pyre. Union interrupted by calculated violence." The words spilled from her in diminishing volume, the vision beginning to release its hold. "Threads cut before their pattern completes. Divine intervention through mortal instruments."

The silver light receded from her eyes in a gradual tide, revealing normal irises beneath as her awareness returned to present reality. Her body sagged against Edrin's support, suddenly boneless with the exhaustion that followed such intense prophetic episodes. Her skin had cooled dramatically, taking on the chill of winter despite the hall's warmth.

"I saw it burning," she whispered against his chest, these words fully her own, stripped of the vision's resonance but carrying the weight of genuine fear. "All of this, consumed. Because of us. Because of our binding."

The hall had emptied of villagers, the rehearsal participants ushered outside by quick-thinking elders who recognized the need for privacy following oracular manifestation. Only Edrin's companions remained—Bran already retrieving his notebook to record details before they faded, Alira approaching with healer's concern evident in her posture, Theryn securing the hall's entrances with quiet efficiency.

"Let's move her to the side chamber," Alira suggested, her hands already preparing a restorative tincture from the components she carried. "She needs stillness to integrate what she's seen."

In the smaller room ordinarily used for elder councils, they gathered around Lysera, who sat wrapped in Edrin's mantle, the woven love glyph positioned over her heart. Color had begun returning to her face, though her hands remained cold enough that Alira chafed them between her own warmer palms while administering sips of herbal infusion designed to ground consciousness firmly in present reality.

"Tell us what you can," Bran prompted gently, notebook open, but his scholarly eagerness was tempered by genuine concern for his friend. "Only what feels clear enough to share."

Lysera's gaze focused on the middle distance, sorting fragmented images into a coherent sequence. "Attack during the ceremony itself. Fire as a both distraction and destruction. They—whoever they are—are using the chaos to reach specific targets." Her eyes lifted to meet Edrin's. "Us. Our binding represents a threat they're specifically tasked with preventing."

"Malrik," Edrin concluded, the name was no longer merely an opponent but an avatar of forces determined to stop what they had begun. The muscle in his jaw tightened visibly beneath his skin. "Or those working under his direction."

"The pattern aligns with his previous incidents," Bran confirmed, scales shifting to troubled bronze as he cross-referenced mental archives. "Significant dates, community gatherings, moments of power alignment—all have attracted intervention when they intersect with our presence." His clawed finger tapped against parchment in unconscious rhythm. "Statistical analysis suggests postponement would be prudent."

Theryn leaned against the doorframe, his posture deceptively casual while his eyes maintained constant assessment of potential threats. "I could arrange traps along likely approach routes," he offered. “Nothing lethal. We need information more than corpses, but sufficient to slow advance and provide warning.”

“The ceremony could be moved,” Alira suggested, her practical nature seeking compromise between celebration and safety. “The old watchtower at the village boundary offers a better defensive position. Limited approaches, stone construction resistant to fire.”

Silence fell as they considered options, the weight of decision settling most heavily on the couple whose union had transformed from personal commitment to potential catalyst for disaster. Lysera’s hand found Edrin’s, their fingers intertwining with unconscious precision, the contact grounding both in something more immediate than strategic calculations.

“How certain is this vision?” Edrin finally asked, his voice steady despite the gravity of the discussion. “Is this inevitable, or one possibility among many?”

Lysera considered the question with Seer’s precision. “Vision shows potential, not certainty. Strong potential in this case. The images carried weight and clarity I’ve rarely experienced.” Her fingers tightened around his. “But threads can be rewoven before they manifest fully into physical reality.”

“Then we prepare for what might come while refusing to surrender what matters,” Edrin stated, decision crystallizing within his words. “We’ve been running and fighting for months. If we postpone living until all threats pass, we may never live at all.”

Lysera straightened, strength returning to her posture as she processed his words. “Staying alive serves no purpose if we don’t enjoy being alive," she agreed, the phrase carrying the weight of personal philosophy rather than mere observation. "We cannot allow fear of what might come to rob us of what we've found."

They gazed at each other with shared understanding—two people who had witnessed enough darkness to recognize light's precious nature, who understood that safety offered no guarantee of fulfillment. The love glyph at Edrin's wrist pulsed gently, responding to the conviction that flowed between them.

"We proceed with the wedding," Lysera concluded, her voice gaining strength with each word. "But we prepare for what my vision revealed. We reweave the pattern to our advantage."

The room's atmosphere shifted subtly—resignation and concern yielding to renewed determination. Bran nodded once, accepting their decision with respect despite his statistical reservations. His scales shifted to deep copper as he immediately began calculating defensive adjustments to ceremonial arrangements.

"The village has three underground storage chambers that could serve as evacuation points," he noted, scholarly focus providing practical response to emotional decision. "I'll modify the elder procession to position the strongest defenders near vulnerable entrances."

Alira moved to the small window, gazing toward the forest edge where a threat potentially gathered. "I'll strengthen the warning wards beyond standard boundaries," she offered. "And prepare emergency supplies—both medical and practical—in accessible caches."

"I'll ensure our unwelcome observers find misleading information," Theryn added, his usual playfulness replaced by professional calculation, though said through a grimace. "False schedules, incorrect participant numbers. And I'll recruit reliable villagers for perimeter watch without raising a general alarm."

Edrin pulled Theryn aside for a moment, "If the worst happens, protect my cousins with your life. Next to you all, they are the only other target that Malrik could use against me. They came to help rebuild this place, and I can't lose them." He immediately left Theryn's side and went to Lysera.

Edrin and Lysera remained connected through their clasped hands; this physical link stayed despite divine opposition. Around them, their companions began transforming decision into action—Bran sketching modified procession routes, Alira inventorying available resources, Theryn slipping away to establish surveillance counterpoints.

"They will come," Lysera said softly, for Edrin alone. "My vision holds too much certainty to ignore completely."

"Then they will find us together," he replied, equal quietness carrying absolute conviction. "As we have faced everything else." His free hand rose to touch her face, thumb tracing the delicate line of her cheekbone with blacksmith's gentleness. "Some fires forge rather than destroy."

Beyond the hall's walls, Emberfield continued its preparations for both Hearthbless and wedding celebration—children practicing songs that would welcome the ceremonial binding, cooks tasting stews and adjusting seasonings for the feast, decorations receiving final adjustments to catch tomorrow's light at optimal angles. The village moved in rhythms established through generations of winter festivals, unaware that these traditions now intersected with cosmic currents of much greater significance—that what appeared as a simple community celebration had become a focal point for forces that operated beyond mortal understanding.

Yet within this intersection of humble tradition and divine conflict, Edrin and Lysera had staked their claim—not for safety or certainty, but for the right to choose connection despite danger, to celebrate finding each other in a world where such discovery represented a miracle worth preserving at any cost.

CHAPTER SIXTEEN
Emberfield

Chapter Sixteen

Emberfield – Morning (Commonsday, 30 Newthread, TT 1215)

The communal hall of Emberfield transformed into a temple of winter light, late afternoon sun angling through frost-kissed windows to paint the floor in amber geometric patterns. Evergreen boughs hung from rafters in precise arcs, their deep green needles intersecting with red ribbons fluttering in the warm currents rising from the central hearth. The ceremonial flame bowl, hammered copper burnished to mirror brightness by Edrin's own hands, waited empty at the circle's heart, surrounded by intricate arrangements of hearth-rings and winter blooms whose placement followed ancient patterns known only to village elders and those who see beyond ordinary sight. Anticipation hummed through the gathered villagers like a struck chord, each face turned toward the hall's opposing entrances where bride and groom would soon appear.

They came as tradition dictated—from separate directions to symbolize distinct lives about to converge. Edrin entered from the western door, the deep indigo of his ceremonial tunic absorbing light while the silver embroidery along its edges caught and reflected it. The fabric, carefully selected and crafted by village hands, softened his warrior's stance without diminishing the essential strength that radiated from his frame. The love glyph at his wrist pulsed with subtle golden light, visible beneath his sleeve cuff as he moved through the assembled villagers. His eyes, dark and steady, searched the room's eastern side where Lysera would appear.

She emerged; the silver-embroidered midnight blue of her robes creating the impression of a night sky brought indoors. Winter bloom flowers formed a delicate crown among her braids, their rare white petals seeming to gather and intensify the room's ambient light. The ceremonial veils floated around her face like morning mist over still water, occasionally parting to reveal the quiet certainty in her expression. Where Edrin moved with the measured precision of a man accustomed to controlling physical force, Lysera glided with the fluid grace of one whose perceptions routinely transcended ordinary boundaries.

They converged at the circle's center, where the village elder—a white-haired woman whose authority resided in both accumulated years and demonstrated wisdom—waited beside the ceremonial flame bowl. Her hands bore the intricate henna patterns specific to Hearthbless, each curve and line representing aspects of community bonds. When she spoke, her voice carried the resonant timbre of one accustomed to being heard and heeded.

"We gather as our ancestors gathered, at winter's heart, to strengthen the flames that sustain us through darkness." Her gnarled fingers gestured toward the empty bowl. "Edrin Ashwood, Lysera Vayesh, you come before this community to join your separate flames into a shared hearth. Step forward and place your hands over the vessel that will contain your union's first fire."

They moved in perfect synchronization, hovering their palms above the copper bowl, not quite touching each other but aligned to allow their fingers to interlace when the moment arrived. The elder reached for the ceremonial striker—a tool of flint and iron that Edrin had forged specifically for this occasion—and prepared to light the first spark that would ignite the oil-soaked kindling waiting in the bowl's center.

The first sign came so subtly that only those with heightened awareness noticed. Throughout the hall, candle flames that had burned with steady golden light suddenly flickered, their color shifting momentarily toward cool blue before stabilizing. Bran's scales rippled in response, darkening to wary bronze as his reptilian eyes narrowed. Near the hall's southern wall, Alira's fingers closed instinctively around her healer's pendant, its flower petals contracting as if sensing atmospheric change.

Edrin and Lysera remained focused on each other, the space between their hands growing charged with anticipation as the elder raised the striker. As she brought the flint and iron together, producing the first spark to ignite their binding flame, the candles throughout the hall simultaneously flared brilliant purple—an impossible color for natural fire.

"By the ancient bonds of hearth and heart, I call forth the flame that—" The elder's voice faltered as her breath suddenly fogged before her face. The temperature in the hall plummeted with supernatural speed, frost crystallizing along window edges and creeping across the floor in delicate fractals that resembled glyphs of warning.

Villagers murmured in confusion, their festival cheer giving way to uncertain glances. Throughout the room, surfaces that should not have been reflective—wooden tables, fabric banners, even the petals of winter blooms—suddenly bore mirror-like qualities. Each surface developed an impossible depth, suggesting windows into other realities rather than simple reflection.

Lysera's expression tightened, the seer's perceptions registering the wrongness permeating their sacred space. "Edrin," she whispered, her eyes meeting his with warning that required no further explanation. His fingers flexed instinctively, reaching for a hammer that wasn't at his side, the warrior's readiness asserting itself beneath the bridegroom's ceremonial attire.

The ceremonial flame bowl between their hands developed a skin of ice across its surface, crackling as it spread to encase the waiting kindling. From this frozen heart, impossibly, rose coiling tendrils of shadow—not mere absence of light but something that possessed mass and intention. These dark manifestations writhed upward, multiplying and expanding until they formed a canopy of living darkness above the central circle.

Glass shattered with explosive force as every window in the hall imploded simultaneously, shards flying inward only to halt mid-air, suspended in perfect stillness before arranging themselves into complex geometric patterns floating throughout the chamber. Villagers cried out, some falling to the ground in protective huddles, others rushing toward exits that seemed much farther away than architectural reality should allow.

Through the main entrance strode Malrik—transformed beyond recognition from the boisterous adventurer he had once been. His skin had become a landscape of deep fissures across his face and body like lightning frozen in flesh. From these cracks glowed the orange-red radiance of molten blood, creating the impression of a human vessel barely containing volcanic force. He wore no clothing save for a kilt-like wrap of material that seemed to shift between solid and smoke, occasionally revealing glimpses of legs partially transmuted into living shadow.

Around his massive form swirled contradictory elements—flames that consumed and shadows that moved with deliberate purpose, these opposing forces creating a corona of power that distorted the very air through which he moved. His dreadlocks, bound with copper wire, glowing white-hot against his scalp, floated upward. Most disturbing were his eyes—once warm brown, now burning with orange flame that occasionally leaked from their corners like molten tears.

"What celebration is this?" Malrik's voice resonated at frequencies that caused suspended glass shards to vibrate in sympathetic harmony. The sound emerged layered, as if multiple voices spoke simultaneously from the same throat. "A binding of powers that challenge divine authority? A union that dares weave patterns not sanctioned by the Loom's keepers? And no invitation for me?"

His gaze fixed on Edrin and Lysera, still standing before the frozen flame bowl, their hands now clasped in protective solidarity rather than ceremonial pose. The shadow-serpents writhing above them seemed to respond to Malrik's attention, their coils tightening in anticipation.

"I bring a wedding gift," he proclaimed, arms spreading wide in a terrible parody of benediction. "The blessing of true sight." With this declaration, he brought his hands together in a thunderous clap that sent a concussive force throughout the hall.

The suspended glass shards multiplied exponentially, each fragment splitting into dozens more until the air itself was composed more of cutting edges than space. These mirror-like splinters rotated in perfect unison, catching and reflecting light not from the hall's interior but from somewhere else entirely.

Villagers screamed as the mirrors revealed personalized horrors—scenes of loved ones burning alive, children consumed by shadow, homes collapsing into endless pits, bodies transforming into twisted abominations. These were not mere illusions but glimpses into potential futures. The shards oriented themselves differently for each viewer, ensuring that none could escape witnessing their particular doom.

"See what comes," Malrik intoned, his fractured face splitting into a terrible smile, revealing teeth partially transmuted into obsidian shards. "See the price of defiance. See the inevitable consequence of challenging gods who were ancient before your ancestors learned to kindle flame."

The mirrors surrounding Edrin and Lysera showed different visions—not personal destruction but wider devastation. Cities consumed by divine fire, mountains reshaping themselves to crush settlements beneath, oceans rising to drown coastal populations, and shadows that devoured entire communities without leaving bodies to bury. Each reflection revealed another facet of the potential catastrophe should they continue their path of questioning divine authority.

Through this crystalline nightmare, Malrik advanced toward the center circle, each footstep leaving behind footprints as if his feet were brands, carving footprints in the wooden floor and leaving them smouldering. Fire and shadow pulsed around him in alternating waves, manifestations of the dual divine powers he channeled through his transformed flesh.

"Your binding ends before it begins," he declared, voice breaking into harmonics that made the air itself tremble. "Your questions end before they find answers. Your defiance ends before it inspires others." His hand extended toward the frozen flame bowl, fingers splayed as fire and shadow gathered at his fingertips, preparing to unleash forces that would transform celebration into annihilation.

#

Panic erupted in waves, rippling outward from Malrik's blazing form, transforming the orderly gathering into fractal patterns of flight and terror. Villagers stumbled over benches and collided with each other as they fled from personal horrors reflected in the floating shards—a mother shielding her child's eyes from visions of familial destruction, an elder falling to his knees before a mirror showing his ancestors rising from graves with accusing fingers, a young couple clutching each other as their reflected selves aged and decayed in seconds. The ceremonial hall, so carefully arranged for celebration, became a labyrinth of fear where exits retreated, and the very air thickened with Malrik's terrible laughter echoing across.

Lysera's training asserted itself through instinct honed by years among the Vayesh seers. Her perceptions identified the architecture of Malrik's attack. Not mere chaos, but calculated devastation designed to fragment communal bonds through individualized terror. The silver flecks in her eyes brightened to metallic brilliance as she channeled her power not into sight but protection.

"Edrin," she called, her voice cutting through the cacophony with preternatural clarity. "Guard my back. I need moments to weave."

Without hesitation, he positioned himself between her and Malrik's advancing form. Meanwhile, Lysera's hands moved in precise gestures to gather invisible threads from the air itself. Her fingers wove complex patterns that left faint silver traces lingering momentarily before dissolving into the hall's charged atmosphere. With each completed sequence, her breathing synchronized with rhythms imperceptible to ordinary senses—the pulse of protection glyphs activated beneath floor and beam during their earlier preparations.

"Na-sheleth, na-voran, na-elith," she chanted, ancient Vayesh words of warding that had protected seers since before written history. From her outstretched hands flowed gossamer strands of silvery light expanding outward, forming translucent barriers between terrified villagers and the most vicious of the floating shards.

"This way!" she called, her voice carrying harmonic undertones that penetrated the confusion clouding panic-stricken minds. "Follow the silver light! The reflections lie—the exits remain where they have always been!"

The protective weaving created pathways of reality through Malrik's illusory maze—corridors of relative safety where the mirror-shards' influence diminished enough for villagers to perceive actual architecture rather than nightmare geography. Children responded first, their less-rigid perceptions more readily accepting the silver guidance, pulling adults along routes that suddenly appeared navigable.

Across the hall, Bran moved with methodical precision that defied the surrounding chaos. His lizardfolk senses processed the disaster through multiple layers of analysis—the physical danger of actual glass, the psychological impact of targeted illusions, the underlying energy patterns that sustained both. From his vest pocket, he withdrew a tuning fork of peculiar design, its tines composed of metal that had never seen daylight before Bran refined it in his laboratory.

Striking it against his scaled palm, he held the vibrating instrument before him like a compass, its resonance revealing invisible currents of power that flowed through the hall. His vertical pupils contracted to thin lines as he followed these patterns in seemingly random trajectories with an underlying purpose.

"Frequency modulation suggests a central control point," he muttered, the scientific precision of his assessment creating stable focus amid emotional chaos. "Convergent pattern implies geometric anchor rather than sustained channeling."

The tuning fork's vibrations intensified as he approached the overturned ceremonial table. Dropping to his knees, Bran examined the floor beneath where the table had stood. Embedded in the wooden planks—glowing with malevolent purpose visible to his specialized perception—lay an obsidian disc inscribed with a mirror glyph of extraordinary complexity.

"Fascinating construction," he observed, scholarly instinct momentarily overriding survival concerns. "Pre-planted focal point activated remotely—efficiency considerably higher than maintaining individual illusions."

From his satchel, Bran extracted a crystal plate of precisely calculated thickness. Positioning it above the hidden glyph, he angled the translucent mineral to refract ambient light into a concentrated beam—a counter-frequency causing the obsidian disc to vibrate at an unstable pitch. The fork in his other hand altered its tone in response, the harmonics between instruments creating dissonance that physically manifested as cracks spreading across the mirror glyph's surface.

"Theoretical application confirmed," Bran noted with satisfaction as he raised his staff—a walking aid that doubled as a research tool and, in moments like this, a weapon. With calculated force, he brought the metal-capped end down upon the crystal plate, driving the concentrated counter-frequency directly into the glyph's heart.

The obsidian disc shattered with concussive force, rippling through the hall's mystical architecture. Throughout the chamber, nearly half the floating mirror-shards disintegrated instantly; their sustaining pattern disrupted at the source. The remaining fragments, however, revealed a crucial distinction. Where the vanished shards had been pure illusion, those that remained possessed physical substance.

A woman near the eastern exit screamed as a mirror-shard cut across her forearm, the wound bleeding actual blood rather than disappearing as illusory injury would have. An elder stumbled as another fragment embedded itself in his shoulder, the pain entirely real despite the phantom nature of what the mirror reflected.

"Physical components remain active!" Bran called out, his voice carrying the precise cadence that maximized acoustic travel through the chaotic soundscape. "Illusion collapsed, but material threat continues!"

Near the hall's northern wall, Theryn's attention fixed on a figure crumpled against a support beam—Maris Ashwood, one of Edrin's distant cousins, a thin line of blood tracking from her temple where a shard had grazed her. Around her, mirror fragments oriented themselves with predatory intelligence, reflecting not random horrors but specifically targeted visions of her cousin's destruction. The mirrors' focus suggested deliberate targeting. Malrik had identified her as leverage against Edrin.

Theryn studied the scene carefully, his hands patting his pouches until he found one of his Threadbraid rollups. He quickly lit it and sucked the smoke into his lungs before moving with the fluid grace that had earned him the title Riddleweaver among his nomadic people. Between heartbeats, he crossed the distance separating him from Maris. His body, trailing sweet-smelling smoke, flowed around floating shards with supernatural precision derived from a lifetime of mastering the art of not being where danger was expected. Reaching her side, he crouched protectively, one arm cradling her semi-conscious form while the other produced a slender blade seemingly from nowhere.

"Stay with me, Cousin-of-Smith," he murmured, using the formal address of respect from his caravan traditions. He breathed out smoke, which seemed to cloud the air between the mirrors. "Edrin battles bravely, and you must witness his triumph."

A particularly large mirror-shard oriented itself toward them, reflecting a vision of Edrin consumed by flames that melted flesh from bone while he screamed for his family's help. Theryn's blade flashed upward, striking the shard at precisely the angle that redirected rather than shattered it. The mirror spun away, its horrific vision now aimed at space rather than vulnerable eyes.

"Clever, sneaky toy soldier," came Malrik's voice from directly behind them, layered with harmonics. "Protecting the leverage. The family-bond—so predictably exploitable."

Theryn turned, positioning himself between Maris and the approaching figure. But instead of Malrik's actual form, they faced a phantom extension that could inflict harm without risking the vessel's safety. The apparition reached toward them with fingers that lengthened into claws of crystallized darkness.

"I know seventeen ways to kill with shadow alone," the projection intoned, its mouth moving in disjointed synchronization with the words. "Shall I demonstrate on the smith's beloved family?"

Theryn's response came not as words but as action. He puffed on the cigar for a moment and then, almost as a secondary thought, his blade described complex patterns that cut the very substance of the projection, disrupting the coherence of its form. Where the metal passed through flame and shadow, momentary vacuums formed, forcing the apparition to constantly reform itself.

"And I know twenty-three ways to disperse what isn't fully material," Theryn countered, the typical playfulness of his speech replaced by cold precision. "Shall I demonstrate on your master's little puppet?"

Across the hall, the combined efforts of Lysera's protective weaving and Bran's disruption of the central glyph had created enough stability for many villagers to reach exits. Alira stood at the eastern door, her fairykin abilities allowing her to identify those most severely injured and prioritize their evacuation. Her hands glowed with healing energy as she pressed fingers against wounds, stabilizing the most critical damage until proper treatment could be administered.

Malrik's attention, divided between multiple targets, suddenly contracted as he registered the failure of his carefully planned trap. The mirror-shards still floating throughout the hall reoriented themselves, abandoning individual targets to form a spiraling vortex around his physical form. Fire and shadow pulsed from the cracks in his skin with increasing intensity, the dual divine energies responding to his mounting frustration.

The phantom attacking Theryn and Maris dissipated as Malrik recalled his extended power, consolidating his focus toward the ceremony's central circle, where Edrin stood guard before Lysera. The love glyph at Edrin's wrist blazed with golden light, extending upward along his arm, forming protective patterns that disrupted the mirror-shards' approach.

"Enough distractions," Malrik declared, his voice causing the very air to vibrate at uncomfortable frequencies. "The architect of defiance stands before me. The questioner of divine authority. The mortal who dares believe he might challenge gods, and the Loom itself."

The swirling vortex of remaining mirror-shards contracted around him like armor—showing the man he had been, others revealing glimpses of the divine forces using him as a vessel, still others displaying something that existed between these states, neither fully human nor truly divine. The composite image created an impression of identity fractured beyond reconstruction, personality shattered and reassembled according to external design rather than internal cohesion.

"Face me directly, Ashwood," Malrik challenged, gesturing with flame-wreathed hand toward the hall's center. "Let us settle what began at Misthold—the question of which vessel carries true purpose."

Edrin's gaze met Lysera's for one suspended moment—silent communication between souls who recognized each other beyond words. Then he stepped forward, moving away from her protective circle toward Malrik's challenge, hammer materializing in his hand from where it had been concealed beneath ceremonial garments.

Behind him, Theryn threw the mantle with Edrin's glyph woven into it over Edrin's shoulders and then returned to Maris. Lysera's protective weaving expanded to encompass the remaining villagers, creating silver barriers that guided them toward safety while her companions regrouped around her. Bran, with his resonating fork, still detected power fluctuations. Alira's hands glowed with healing energy prepared for whatever injuries might come. Their coordinated efforts had transformed Malrik's carefully orchestrated trap into something approaching stalemate—survival with purpose intact.

#

The communal hall bore little resemblance to the sacred space that had, mere minutes before, awaited the binding of two souls in ceremonial union. Shattered glass crunched beneath Edrin's boots as he advanced toward the center, where Malrik stood amidst swirling fragments of mirror and shadow. Overturned benches formed a crude arena boundary, while tattered evergreen boughs hung from rafters like the broken wings of defeated birds. The ceremonial flame bowl lay twisted and cooling several paces away, its purpose corrupted before fulfillment. Through gaps in the damaged roof, winter sky peered down with cold indifference—stars emerging in deepening twilight to witness the confrontation between mortal determination and divine proxy.

Edrin halted ten paces from his adversary, his ceremonial attire torn and smudged yet somehow lending dignity rather than diminishing it. The hammer in his hand—his personal blacksmith's tool retrieved from its hiding place—hung at his side with the comfortable weight of extension rather than burden. His eyes, level and unwavering, met Malrik's flame-filled gaze across the distance that separated man from whatever Malrik had become.

"No gods," Edrin stated, each word precise and measured. "No mantles. Just you and me, Malrik." He raised the hammer, not in threat but presentation. "Steel against steel. Skill against skill. Man against man." His voice carried neither fear nor bravado but something rarer—certainty born from self-knowledge rather than delusion. "Or has whatever remains of you become too afraid to face me without borrowed power?"

Laughter erupted from Malrik's fractured face, the sound containing deep resonances that vibrated floorboards and high harmonics that caused remaining glass to sing in sympathetic pitch. Flames licked between his teeth as he spoke, casting his features in hellish relief against the gathering darkness.

"Afraid?" The word emerged distorted, stretched across multiple octaves simultaneously. "Your challenge amuses the vessel and offends the power it contains." Malrik spread his arms, the gesture causing cracks across his chest to widen, revealing glimpses of molten interior that pulsed with unnatural rhythm. "I am beyond man, beyond mortality's constraints. Why would divinity stoop to such limitations?"

"Because you can't help yourself," Edrin replied, the quiet confidence in his tone a striking counterpoint to Malrik's theatrical display. "You dared to interrupt my life! I will show you why that was a mistake. And, the man who sought glory through adventure still hungers for recognition. Even as a second-rate divine vessel, something in you needs to prove your worth. Down deep, in the bones, *you* want to beat me. Pyrrhos could care less, because Pyrrhos could cut me down and I couldn't stop it, but you...."

Something flickered in Malrik's burning eyes—a momentary contraction suggesting Edrin's assessment had struck something human beneath divine overlay. Without warning, his arm snapped forward, palm extending toward Edrin as a perfect sphere of shadow-fire materialized and launched across the intervening space.

Edrin dove sideways, the projectile passing so close that he felt simultaneous sensations of scorching heat and absolute cold against his cheek. The shadow-fire struck a support beam behind him, the wood simultaneously blackening to ash and crystallizing into obsidian-like material before collapsing in brittle chunks.

"Your mortal reflexes won't save you indefinitely," Malrik observed, his massive form already circling to maintain optimal striking distance. His movements contained bizarre duality—his upper body flowing with flame's fluid unpredictability while his legs advanced with the ponderous inevitability of geological processes. "Each dodge merely delays the inevitable consumption."

Edrin regained his footing, hammer raised defensively as he matched Malrik's circular trajectory in the opposite direction. His breath came steady and controlled, the discipline of forge-work evident in how he conserved energy while assessing his opponent. The love glyph at his wrist pulsed with increasing brightness, its golden light extending tendrils up his forearm in a protective lattice.

"You're wrong about inevitability," Edrin countered, eyes cataloging subtle details of Malrik's transformed state. The cracks in his opponent's skin weren't static but dynamic—widening and narrowing with the pattern revealing a rhythm that suggested potential vulnerability. "Nothing's inevitable while choice remains."

As they circled one another, the deterioration in Malrik's physical form became increasingly apparent. What had initially appeared as a controlled transformation now revealed signs of structural failure—the fissures across his skin deepened, molten substance occasionally dripping from widening gaps to sizzle against the wooden floor. The dual mantling created visible conflict within his flesh, areas influenced by Pyrrhos glowing with increasing heat while regions dominated by Umbryos' shadow grew unnaturally dark, the boundary between these territories shifting constantly across his torso.

Malrik's eyes narrowed as he registered Edrin's analytical assessment. "You see decay where there is ascension," he snarled, the sound accompanied by gouts of flame erupting from his mouth and nostrils. "The vessel adapts to contain divinity's fullness." His hand rose to press against his chest where the largest fissure pulsed with a particularly violent rhythm. "Mortal limitation burns away to make room for perfection."

"I see a man burning up from within," Edrin replied, continuing his measured circuit around the hall's center. "Not transcending mortality but being destroyed by forces he doesn't understand." He gestured with his free hand toward the molten drops marking Malrik's path across the floor. "Gods don't bleed, Malrik. Even divine fire."

Rage contorted Malrik's features, the emotion disturbing the precarious balance of energies within him. Shadow surged momentarily dominant along his right side, causing that portion of his body to blur at its edges, losing definition. He raised both hands, flame gathering around one while darkness condensed around the other.

"Your pathetic mortal limitations blind you to the glory of transformation," he growled, the words emerging strained. "You could have had this! But you cling to flesh's imperfections, to community's weakness, to love's temporary comfort—all shadows of true power, all substitutes for divinity's perfect certainty."

Three shadow-fire projectiles launched in rapid succession—not directly at Edrin but at points that formed a triangle around him, limiting potential evasion routes. Edrin responded not with backward retreat but forward advance, rolling beneath the lowest projection and coming up within striking distance. His hammer swept in a horizontal arc toward Malrik's knee to test his defensive capabilities.

Malrik's leg transmuted to shadow-substance at the moment of impact, the hammer passing through with only minimal resistance. Yet the conversion wasn't completely controlled. The shadow-state lingered longer than intended, forcing Malrik to rebalance on his solid leg while struggling to re-materialize the affected limb.

"Impressive, Malrik," Edrin acknowledged, already dancing backward beyond immediate counterattack range. "But costly. How much concentration does each change demand? How many can you maintain simultaneously before your focus splinters?"

The questions weren't merely tactical banter but calculated psychological assault—each prompting Malrik to perform mental calculations that further strained his already overtaxed cognitive processes. Divine power flowing through the mortal mind created constant pressure, each additional task increasing the likelihood of processing failure.

Malrik launched himself forward with explosive speed, his upper body trailing flame like a comet's tail while his lower half left footprints of absolute darkness lingering momentarily before dissipating. His arms extended into elongated forms that resembled claws of crystallized fire, slashing at Edrin in patterns that defied anatomical possibility.

Edrin met the assault with blacksmith's economy of movement—no wasted motion, no unnecessary exposure. Where flame-claws swept toward his face, his hammer intercepted not to block but to deflect, changing the attack's vector just enough to convert a lethal strike into a near-miss. When shadow-tendrils attempted to encircle his legs, he stepped precisely at angles to disrupt their cohesion. Throughout, his awareness remained multifaceted—tracking Malrik's deteriorating physical condition even while defending against supernatural assault. Edrin noted something and smirked, "I wonder, Malrik, how many appendages you can shadow?" And he swiped the hammer in a fake attack to the groin area, which was completely human. Malrik immediately jumped back to protect.

The confrontation escalated in intensity as Malrik grew increasingly frustrated by his inability to land a decisive blow. His attacks became less coordinated, raw power compensating for diminishing finesse. The fissures across his body widened alarmingly, molten substance flowing more freely from deepening cracks. His breathing emerged labored and irregular, each exhalation accompanied by clouds of smoke and shadow.

"You cannot…" Malrik gasped, momentarily hunching as a particularly violent energy surge caused new fractures to spider across his shoulder and neck. "You will not deny the gods their rightful…"

The sentence remained unfinished as sudden disruption rippled through the hall's mystical architecture—the moment when Bran shattered the hidden mirror glyph beneath the ceremonial table. The effect on Malrik was immediate and profound. His body convulsed, the carefully maintained balance between fire and shadow temporarily collapsing. For one crucial moment, his defenses faltered completely as his consciousness scrambled to reestablish control over competing divine energies.

Edrin recognized the opportunity with warrior's instinct. In the suspended heartbeat between Malrik's vulnerability and recovery, he closed the distance with explosive acceleration. The hammer in his hand described a perfect arc born from thousands of hours at the forge—not a wild swing but a precisely calculated strike targeting the largest fissure across Malrik's chest where molten blood pulsed most visibly.

The impact resonated through the hall with a sound like a mountain splitting. Malrik's body absorbed the blow's kinetic energy before translating it into catastrophic expansion of the targeted fissure. The crack widened dramatically, molten interior substance erupting outward in pressurized spray that carved a glowing parabola through the air. Genuine pain—perhaps the first authentic sensation he had experienced since transformation began—distorted Malrik's features into a mask of shocked disbelief.

"Impossible," he gasped, one hand clutching the wound where the hammer had struck true. The fiery glow in his eyes flickered momentarily, revealing glimpses of human irises beneath divine overlay. "Mortal steel cannot harm divine instrument..."

He staggered backward, leaving a trail of molten droplets that sizzled against the wooden floor. The corona of fire and shadow surrounding him contracted erratically, portions winking out entirely before resurging in uncoordinated bursts. For the first time since his dramatic entrance, uncertainty replaced arrogance in his expression—the realization that his assumed invulnerability had been proven false, introducing an element of fear previously absent from his calculations.

"Gods may not bleed," Edrin replied, hammer still raised in readiness for continued conflict, "but men always do, even when they channel divine power."

Malrik's features contorted with emotions too complex and contradictory for simple naming—rage and fear, pain and pride, determination and doubt all warring for dominance across the fractured landscape of his face. The internal conflict manifested physically as flames surged along one side of his body while shadow deepened along the other, the boundary between these influences shifting unpredictably across his torso.

"This changes nothing," he snarled, though the declaration's impact was undermined by his continued backward retreat. "The vessel may require… temporary restoration. But divine purpose remains unaltered." His form began dissolving at its edges—not a controlled transformation but an emergency evacuation as the gods reclaimed their power from damaged housing.

"Tell your masters - that mirror witch, and your gods - I'll face them too," Edrin called as Malrik's figure became increasingly indistinct, smoke and flame obscuring his retreating form. "No matter what vessel they choose, no matter what power they wield—some questions deserve answers, even if gods fear giving them."

Malrik's final response came as a wordless howl, causing dust and splinters to rain from damaged rafters. Then he was gone—vanished in an explosive surge of smoke and shadow that left behind scorched floorboards and lingering scent of sulfur. The departure carried none of the ceremonial grandeur of his arrival—not victory's triumphant exit but retreat's hasty withdrawal.

In the sudden silence, Edrin stood alone at the hall's center, hammer lowered to his side as he surveyed the destruction surrounding him. The love glyph at his wrist pulsed once more before settling into a steady, subtle glow—not extinguished by conflict but tempered by it.

#

Twilight deepened into true night as the last echoes of battle faded from Emberfield's communal hall. Candlelight—ordinary flame restored to natural gold after supernatural corruption—cast gentle illumination over a scene more suited to the aftermath of war than celebration. Broken glass littered the floor in constellations of danger, catching light in glittering patterns. The wooden beams overhead, scored by shadow-fire and weakened by structural damage, groaned occasionally. Throughout the space, villagers moved with the numb precision of those navigating boundaries between shock and necessary action—tending wounds, gathering salvageable materials, rebuilding what small portions of normalcy remained within reach. They worked with the determination of those who had done it once before and would rebuild again.

Edrin knelt beside an elder whose arm bore the laceration of a mirror-shard, applying pressure to the wound while Lysera prepared a binding cloth. Their ceremonial attire hung in tatters, yet they worked with calm efficiency that provided an anchor for those around them. Though physically separated by their duties, an invisible current flowed between them—shared purpose creating a connection that transcended mere proximity. When their eyes occasionally met across the wounded, an unspoken understanding passed between them, grief acknowledged but not permitted dominance over necessity.

"The bleeding's slowing," Edrin murmured, lifting the compress slightly to assess the injury. "Clean edges—it should heal well if we prevent infection."

Lysera nodded, her fingers moving with seer's precision as she applied healing salve to the bandage. "The physical wounds concern me less than the mental ones," she replied softly, gaze shifting momentarily toward a young girl who sat staring at nothing, hands trembling from visions the mirrors had shown her. "Some scars form beyond flesh's boundaries."

Across the hall, Bran crouched among scattered mirror fragments, his scaled hands moving with meticulous care as he collected specific shards for examination. His spectacles reflected candlelight in twin discs that obscured his reptilian eyes, giving the impression of a scientist rather than warrior—knowledge-seeker amid the battlefield. Each fragment he selected went into separate cloth pouches, labeled according to a classification system comprehensible only to him. Occasionally, his tuning fork emerged to test a particular shard's resonance frequency, the data carefully noted in his ever-present journal.

When he finally approached Edrin, Bran's scales had settled into the deep bronze, signaling serious concern beneath his scholarly exterior. He waited until the wounded elder had been helped to a recovery space before speaking, his voice pitched low to prevent unnecessary spread of disturbing information.

"What we witnessed confirms theoretical concerns about Malrik's condition," he began, producing a particularly large mirror fragment from his collection. The shard's surface reflected swirling patterns of fire and shadow that moved independently of physical surroundings. "Two distinct divine influences occupying a single mortal vessel—Pyrrhos and Umbryos simultaneously."

Edrin's expression darkened as he wiped blood from his hands with a cloth Lysera provided. "I've never heard of anyone channeling two gods before. Is that even possible?"

"Not sustainably," Bran replied, his clawed finger tracing the boundary where fire met shadow within the mirror's reflection. "The human form—even one specially prepared through ritual and sacrifice—lacks sufficient capacity to maintain distinct divine energies in stable configuration. Note the progressive deterioration of his physical integrity during confrontation."

Lysera joined them, her hand finding Edrin's unconsciously, a gesture that had become habit in moments of shared concern. "The cracks in his skin were widening throughout the battle," she observed. "And his control seemed increasingly erratic."

"Precisely," Bran confirmed, his scholarly satisfaction evident. "What we witnessed represents unprecedented manipulation of divine power—and fatal miscalculation for the vessel." His voice lowered further, forcing them to lean closer. "But most concerning is what these mirror fragments reveal about the source of manipulation."

He produced a second, smaller shard, more intricately patterned. When turned at a specific angle, this fragment revealed a brief glimpse of another figure. A female form with fractured eyes like embedded mirror shards, her presence visible for a mere heartbeat before dissolving back into swirling elements.

"The mirror witch," Edrin stated, recognition immediate and grim.

"Indeed." Bran's vertical pupils contracted to thin lines as he examined the fragment. "These mirror sigils bear her distinct resonance pattern. But her role extends beyond mere enchantment." He placed both fragments side by side on a relatively clean section of the floor. "My analysis indicates she's not merely controlling Malrik—she's using him as a conduit."

The term hung between them, its significance not immediately apparent to non-scholars. Bran continued, shifting into the precise cadence he adopted when explaining complex concepts.

"Traditional divine mantling involves a direct connection between god and chosen mortal—power flowing from divine source to human vessel," he explained. "What the mirror witch has engineered is fundamentally different. She's positioned herself as intermediary, a third point in what should be a linear connection."

Understanding dawned in Lysera's expression, her seer's training allowing quicker comprehension of mystical mechanics. "She's siphoning some of the power," she murmured, fingers unconsciously tracing the protective glyph at her wrist. "Drawing it through Malrik rather than directly from the gods."

"Correct." Bran's silver tail twitched with agitation rare for his usually controlled demeanor. "She receives a portion of both divine energies without suffering consequences of direct mantling. While Malrik's body deteriorates from the strain of containing Pyrrhos and Umbryos simultaneously, she harvests benefits with none of the physiological costs."

Edrin's brow furrowed as he processed this information. "So the molten blood, the cracking skin—"

"Symptoms of imminent catastrophic failure," Bran confirmed. "Human physiology cannot indefinitely sustain dual divine resonance. The competing energies of fire and shadow create fundamental incompatibility at the cellular level. Eventually, his body will literally tear itself apart as these forces seek separation."

"But why would the gods allow this?" Edrin asked, frustration evident in his tone. "Surely Pyrrhos and Umbryos must realize they're being manipulated."

Bran's scales darkened to troubled copper. "Divine perception operates differently from mortal understanding. It's entirely possible each god believes they maintain an exclusive connection to Malrik, the mirror witch's interference creating a perceptual blind spot through mirror manipulation." His clawed finger tapped against the fragments. "Additionally, gods typically perceive mortals as interchangeable components rather than individuals. The specific identity of their vessel concerns them less than the successful execution of divine will."

"She's killing him," Lysera stated, the blunt assessment cutting through theoretical discussion. "Using him as a sacrificial element in whatever greater working she's planning."

"Precisely." Bran gathered the fragments, returning them to their respective pouches. "And based on progressive deterioration rate, his remaining functional lifespan can be measured in weeks rather than months—assuming he doesn't encounter another hammer blow like today's."

Around them, the hall had gradually transformed from chaos to wounded order. Alira moved among the injured, her fairykin healing abilities focused on those most seriously harmed. Theryn had organized younger villagers into cleanup teams, his natural leadership temporarily overriding his preference for operating alone. The shattered flame bowl had been carefully collected, its twisted remains placed upon cloth at the hall's center—a broken symbol of interrupted ceremony that nonetheless retained totemic significance.

As night fully claimed the sky beyond damaged walls, lanterns and candles created islands of warm light throughout the space. Villagers had begun returning to their homes or gathering in small groups for comfort and security. The immediate crisis had passed, leaving in its wake the slower emergency of recovery and rebuilding.

The companions gathered near the ceremonial space where, hours earlier, a wedding should have united two souls in community witness. Their physical exhaustion manifested in various ways—Alira's usual luminescence dimmed to barely perceptible glow, Theryn's posture lacked its characteristic fluid grace, and Bran's scales settled into dull bronze, suggesting cognitive fatigue alongside physical depletion. Yet beneath these signs of weariness lay something stronger—determination tempered rather than broken by conflict.

Edrin stood at the circle's center where the flame bowl had rested, Lysera beside him, their hands joined. The love glyph at his wrist pulsed faintly, its golden light visible through the torn fabric of his ceremonial sleeve. When he finally spoke, his voice carried the quiet authority that had gradually become his natural register.

"They may have taken our ceremony," he said, gaze moving from Lysera to each companion in turn, "but they can't take what binds us." His free hand rose to touch the glyph at his wrist, an acknowledgment of connection, transcending ritual formalization. "What matters isn't words spoken before witnesses or symbols exchanged in tradition's name. What matters is the choice we make each day to stand together against whatever comes."

Lysera's fingers tightened around his, silver flecks in her eyes catching lantern light. "The Loom recognized our binding before any elder could pronounce it," she agreed, voice soft yet carrying to all who needed to hear. "No divine interference can unweave what has already been woven into reality's fabric."

"Most romantic declaration I've heard today," Theryn quipped, genuine affection softening the words. "Particularly considering the competition involved actual fire and brimstone."

"Statistical analysis suggests relationships tested by crisis demonstrate thirty-seven percent greater resilience than those formed under ideal conditions," Bran observed, scholarly precision providing emotional support in the only way he fully understood how to offer.

Alira stepped forward, her hands still faintly glowing from healing work. "What happened today reveals how deeply they fear what you represent," she said, addressing Edrin and Lysera but including all present in her gaze. "Not just individually, but the pattern you create together—the questions you've begun asking, the answers you're getting closer to finding."

"The pattern we all create together. And questions about who controls the Loom," Edrin nodded, his expression hardening with renewed determination. "About why gods fear mortal understanding of divine mechanics."

"And who stands behind Malrik's transformation," Lysera added, "directing divine power toward a purpose the gods themselves might not fully comprehend."

Their words formed an implicit pact between them—commitment not just to survival but to the pursuit of knowledge that powerful forces clearly wished to prevent them from acquiring. As they stood in a circle of broken celebration, surrounded by evidence of divine opposition, their resolve crystallized into something stronger than before the attack—determination forged by confrontation rather than abstract threat.

Outside, Emberfield settled into an uneasy night, and villagers started securing homes against dangers both seen and unseen. Within the damaged hall, five companions formed a different kind of security. Though no elder had pronounced words of binding, though no ceremonial flame had sealed their commitment, something equally powerful had emerged from the day's devastation. An unshakable conviction that together they would face whatever came next, whether wielded by mortal hands or divine will.

EPILOGUE

Beneath Threadmere

EPILOGUE

Emberfield – Dawn (Thalday, 1 Concord Days, TT 1215)

Dawn crept across the ruins of Emberfield's communal hall as a revealer of wounds. Its pale fingers pulled back night's merciful shroud to expose the full measure of destruction. The makeshift shrine had been assembled from fragments of celebration—twisted metal from the ceremonial flame bowl, charred remnants of evergreen garlands, scorched ribbons that had once symbolized union now woven into patterns of remembrance. Morning frost glittered upon these broken things, transforming devastation into delicate beauty, emphasizing the violence that had created it.

The companions gathered in a crescent before the shrine as villagers moved like shadows at the periphery, their breath forming clouds that lingered in the winter air like unspoken lamentations. No formal announcement had summoned them; grief possessed its own gravity, drawing the survivors together at first light to acknowledge what had been lost and what remained.

Edrin approached the shrine with his father's blacksmith hammer cradled in calloused hands. The tool—worn smooth from generations of use, its handle darkened by the oils of those who had wielded it before him—had survived when newer, more ornamental pieces had been destroyed. His hands trembled visibly as he knelt to place it among the memorial tokens, his fingers reluctant to release this connection to ancestry and craft. The hammer represented continuity through destruction, heritage preserved despite the gods' interference—a statement made in iron rather than words.

"From the forge of my father, and his father before him," he murmured, the traditional phrase emerging rough-edged and incomplete. His throat worked against emotion that threatened to close it entirely. "May your spirits find the pattern... find the path…" The familiar benediction faltered as memories surged unbidden—his father's hands guiding his own at the anvil, teaching rhythm and restraint, the lessons of metal and flame that had shaped him into the man who now knelt amid the aftermath.

Lysera stood beside him, close enough that the hem of her midnight blue robe occasionally brushed his shoulder in silent reminder of her presence. The silver in her eyes had not fully receded since her vision during the ceremony, as if some door between ordinary sight and seer's perception had been forced ajar by yesterday's events and refused to close completely. Through this expanded awareness, she perceived not just physical devastation but wounds in the fabric of community itself—trust fractured, certainty shattered, faith questioned. Her hand descended gently to Edrin's arm, steadying his tremors with a touch that carried reassurance through skin to spirit.

"They hear you," she said softly. "Those lost and those who came before. The pattern holds even when we cannot see its full design." The words emerged with subtle resonance, suggesting both personal conviction and oracular insight.

Behind them, Bran unfolded a wedding banner salvaged from the hall's eastern wall, its fabric scorched around edges that had curled inward like wounded things seeking protection. With methodical care that defied the emotion evident in his shifting scale colors—deep bronze fading to somber gray and back again—he draped the banner over a small cenotaph constructed from fallen roof beams. The careful precision of his movements transformed practical action into ritual, each fold and adjustment creating a physical manifestation of internal order imposed upon external chaos.

When the banner lay arranged to his satisfaction, Bran extracted his memory tome from its protective satchel. The book—bound in leather that had absorbed years of handling into supple flexibility—opened to pages already prepared with careful columns for recording. His clawed finger moved down the list of names, each representing a life disrupted or ended by yesterday's violence.

"Archival accuracy honors their sacrifice," he murmured, more to himself than those around him, his academic cadence providing emotional distance that allowed him to function amid grief. "Sixteen injured requiring significant treatment. Four... four beyond healing." His scales darkened momentarily as he inscribed these numbers.

Alira knelt nearby, her fairykin form seeming both substantial and ethereal in dawn's strengthening light. Her hands, capable of coaxing healing from most grievous wounds, moved with practiced precision among memorial garlands, applying salves whose scent carried hints of maryvine for remembrance, lorach for peace, winterbane for protection against further harm. These were not mere symbolic gestures but practical magic—subtle enchantments woven through plant matter and intention to ease spirits' passage and comfort those left behind.

"The body remembers what the mind cannot express," she whispered to each garland as she completed its treatment, the healer's wisdom acknowledging grief's physical manifestations alongside its emotional weight. "May these carry what words fail to capture."

At the perimeter of this impromptu ceremony, Theryn maintained his vigilant watch, his lean form a sentinel against horizons that had proven treacherous. His usual fluid grace had reconfigured itself into alert stillness, a predator's patience containing potential for explosive movement should threat materialize. His green-gold eyes scanned forest edges and village approaches with methodical thoroughness, yet occasionally returned to rest upon his companions with fierce intensity that revealed protective instinct seldom displayed so openly.

Each sweep of his gaze cataloged potential dangers, calculated defensive positions, mapped escape routes—not from paranoia but from hard-earned understanding that mourning created vulnerability even gods might exploit. His hand rested upon his blade not as a threat but as a promise—that this circle of grief would remain inviolate, that no further harm would reach those under his watch while wounds remained so raw.

Villagers approached in small groups, each bearing tokens that represented individual connections to what had been lost. An elderly woman placed a wooden spoon carved with protection glyphs beside the hammer, her gnarled fingers lingering on both implements. A father guided his young daughter forward, helping her position a crudely fashioned doll where it could 'watch over' the memorial. A youth barely into adulthood added a handful of copper nails—apprentice work, imperfect but earnest—to the growing collection.

With each addition, the shrine transformed from hastily assembled marker to true communal focus—a physical manifestation of shared loss and collective resilience. Their whispered prayers created a subtle auditory texture that wove through the morning air, petitions to various gods forming harmonies and occasional dissonances that reflected Emberfield's theological diversity. Some invoked Morthos to guide departed souls, others petitioned Thaliora for cleansing tears, still others whispered to Gaelora for creative rebirth from destruction's ashes.

The circle widened as more villagers arrived, surrounding the companions in expanding rings of shared mourning. No formal boundaries separated those who had faced divine proxy directly from those who had witnessed the aftermath, yet natural spacing emerged—acknowledgment of the central five as focal points around which community grief organized itself. The pattern they created together—mourners arranged in concentric circles, tokens forming a spiral at the center, dawn light illuminating all from an eastern angle—resembled an ancient glyph of renewal, though none had consciously designed it so.

As morning light strengthened from tentative promise to established presence, silence gradually descended upon the gathering—not empty absence of sound but full-bodied quiet that contained multitudes. Within this shared stillness, individual griefs maintained their distinct characters while participating in a greater whole, like separate instruments resting between movements of a complex composition. The moment stretched, expanded, encompassed them all in temporary suspension between acknowledgment of what had been lost and recognition of what remained to be protected.

When the breeze finally stirred the charred banner, setting its edges fluttering with quiet insistence, the spell of absolute stillness broke into a gentle resumption of necessary movement. The memorial had been established, the fallen acknowledged, the survivors counted among each other's remaining blessings. Now came the harder work of continuing forward while carrying both their dead and their questions toward whatever answers awaited beyond morning's revelation of both destruction and endurance.

#

The group withdrew to a secluded corner of the shrine grounds where ancient pines formed a natural barrier against curious eyes and attentive ears. Morning light filtered through evergreen boughs in dappled patterns that seemed to write messages upon the undisturbed snow. Here, partially sheltered from both elements and observation, the five companions formed a smaller circle within the larger pattern of community mourning, their proximity creating a pocket of shared warmth against winter's persistent chill.

Edrin stood at the circle's center, his ceremonial clothing replaced by simpler attire that bore subtle evidence of hasty repair—stitches slightly uneven where tears had been mended, a charred edge trimmed away rather than restored. Such imperfections seemed appropriate, even necessary, within the context of their interrupted celebration. His eyes, shadowed by exhaustion yet bright with purpose, found Lysera's as he extended his hands toward her in a gesture that mimicked their positions before the ceremonial flame bowl.

"Come," he said simply, the single word carrying weight beyond its brevity.

Lysera moved forward without hesitation, her burnished skin absorbing morning light while the silver in her eyes reflected it with subtle intensity. Her hands met his in perfect alignment, fingers interlocking with familiar precision that required no conscious thought. At their point of contact, the love glyph etched upon Edrin's wrist pulsed with golden radiance that traveled upward along his forearm in delicate tracery—not the dramatic flare of battle-summoned power but steadier illumination suggesting foundation rather than weapon.

"They tried to take this from us," Edrin said, his voice rough with emotion held firmly in check. The muscles along his jaw tightened momentarily before relaxing into deliberate calm. "The ceremony, the witnesses, the community blessing—they sought to deny us everything that formalizes commitment." His thumb traced a small circle against her palm, the intimate gesture creating counterpoint to his words' gravity. "But I won't let them. What lies between us belongs to us alone, not to rituals or traditions, certainly not to gods who fear what we might become together."

Lysera met his gaze with unwavering steadiness, the silver flecks in her eyes catching light in patterns that resembled constellations against a midnight sky. "What we have cannot be broken by fire or blade," she replied, her voice carrying harmonic undertones that belonged to both woman and oracle—personal conviction amplified by seer's certainty. "The glyph manifested from the Loom itself, called forth by recognition of pattern already woven through our separate threads. No power can unravel what the fabric of reality has acknowledged."

Their fingers intertwined more securely, the physical contact mirroring a metaphysical connection that had established itself beyond ceremonial confirmation. The love glyph's radiance stabilized into steady illumination that cast their joined hands in gentle gold, turning simple flesh contact into a symbol containing complex meaning—defiance and tenderness, protection and vulnerability, question and answer existing simultaneously within a single gesture.

Bran stepped forward from his position at the circle's edge, his violet reptilian eyes narrowed behind spectacles that had acquired a faint crack across one lens during the previous day's conflict. From his ever-present satchel, he withdrew his almanac—not the formal memory tome used for official recording but a personal journal where calculations too speculative or unproven for formal documentation took shape. The pages he displayed contained intersecting lines forming a complex grid overlaid with celestial markers and temporal notations.

"We need to understand their timing," he stated, clawed finger tracing pattern across the parchment with scholarly precision that transformed emotional situation into analytical problem. "Divine intervention follows discernible patterns despite the appearance of random malevolence. I've begun tracking convergences between celestial positions and terrestrial power concentrations." His scales shifted to intense copper that signaled cognitive engagement, overriding emotional response. "Note three previous attacks—Misthold, Widow's Fork, and yesterday's interruption—each occurred during a significant astrological alignment that amplified specific divine influences."

The almanac pages revealed careful documentation beyond mere astronomical recording—notations of weather conditions, emotional states, community activities, and magical manifestations arranged in parallel columns suggesting correlation rather than coincidence. Lines connected seemingly disparate elements into an emergent pattern that remained incomplete yet hinted at an underlying order not immediately apparent to casual observation.

"I'll track every convergence, every pattern," Bran continued, his tail twitching with unconscious emphasis. "Statistical analysis suggests a ninety-three percent probability that future divine intervention will coincide with similar alignments. Forewarning provides tactical advantage even against a superior force."

Alira approached next, her fairykin grace temporarily subdued by grief, yet evident in the economy of movement that wasted no energy on unnecessary gesture. From the embroidered pouch at her waist, she withdrew small bundles wrapped in linen squares—each package secured with thread dyed specific colors and knotted in patterns containing meaning beyond mere function. The contents produced a subtle aroma that combined herbs traditionally associated with wedding celebrations—now repurposed through intention and fairykin craft.

"These were gathered for blessing your union," she explained, distributing bundles to each companion with deliberate care, with proper alignment of knot patterns. "When plans shift, wise ones adjust rather than abandon. I've transformed celebration herbs into protection charms through binding and blessing."

As each bundle changed hands, its scent intensified momentarily before settling into a subtle presence, lingering at the edge of conscious perception—not intrusive but a persistent reminder of purpose. Alira's fingers lingered briefly against each recipient's palm, the contact transferring a subtle energy charge connecting separate charms into a cohesive network.

"We carry what was lost," she said firmly, fairykin pragmatism providing a framework for emotional resilience. "Not as burden but as resource—memory transformed into preparation, celebration redirected toward survival."

Theryn accepted his bundle with uncharacteristic solemnity, his usual sardonic expression replaced by focused intensity rarely displayed so openly. He tucked the charm into the inner pocket positioned directly over his heart, the placement deliberate rather than arbitrary. When he spoke, his voice lacked the playful cadence that typically masked deeper considerations beneath surface humor.

"Next time, we'll be ready," he stated, the simple declaration carrying the weight of a complex promise. His green-gold eyes held unusual clarity—the mischief that typically danced within them was temporarily set aside in favor of absolute commitment. "They showed their newest methods yesterday. Revealed tactics they believed would destroy us." One corner of his mouth lifted in an expression too grim to qualify as a smile. "Instead, they provided education we could not have gained through any other means."

Without discussion or signal, the five shifted position until they formed a perfect circle, their bodies angled inward to create shared space at the center. Each extended hand toward this common point, palms facing inward rather than upward or downward. Their fingertips met in gentle contact, physical touch manifesting invisible links that had formed between them through shared experience.

No words accompanied this gesture; none were necessary. The silence between them carried content beyond verbal expression—acknowledgment of individual strengths contributing to collective resilience, recognition of separate pasts converging toward a shared future, acceptance of both vulnerability and power inherent in choosing connection over isolation. What flowed between their joined hands contained questions and commitment in equal measure—determination to discover truth regardless of divine opposition, promise to protect one another through whatever consequences such discovery might trigger.

The love glyph at Edrin's wrist pulsed once more, its golden light expanding outward to illuminate their gathered hands in a brief corona that then subsided into a steady, subtle glow. In that momentary illumination, each face revealed essential nature stripped of pretense or performance—Edrin's quiet strength, Lysera's far-seeing wisdom, Bran's analytical courage, Alira's compassionate pragmatism, Theryn's cunning loyalty. Five distinct individuals whose connections formed a pattern greater than the mere sum of components.

When they finally lowered their hands and stepped back, something had solidified between them—not a formal oath spoken aloud for witnesses to validate, but a deeper commitment forged through shared vulnerability and collective purpose. Their bond, tested by divine fire rather than consecrated by ceremonial flame, had emerged stronger than before the attack.

The silent vow they had exchanged required no external confirmation, no elder's blessing, no community witness. It existed as reality already acknowledged by the Loom itself, threads interwoven with deliberate intention that would not easily unravel, even should gods themselves attempt to cut the pattern before its completion.

#

The cobblestoned village square that bordered the shrine grounds had transformed from a solemn gathering place to something altogether different as mid-morning sun burned away the last tendrils of frost. Tables that had been prepared for the wedding feast stood in long rows, their surfaces crowded with covered dishes that had been stored in root cellars and pantries far from the communal hall's destruction. Steam rose from pots kept warm over carefully tended coals, carrying aromas that spoke of life's persistence—roasted meats rubbed with precious winter spices, vegetables preserved from autumn's harvest, breads whose crusts had been scored with traditional symbols of union and continuity. The untouched abundance stood in stark contrast to the nearby shrine, as if two separate realities occupied adjacent spaces.

Villagers drifted between these territories with uncertain steps, their expressions caught between remembrance and hunger, grief and need. Children stared longingly at honey cakes stacked in golden towers, while adults exchanged glances that contained questions none seemed willing to voice aloud. The feast had been prepared for celebration, now denied them; consuming it felt somehow improper, yet waste amid winter scarcity represented its own kind of wrongness.

Theryn observed this hesitation from the edge of the square, his keen eyes noting the subtle body language, revealing conflict between practical necessity and emotional appropriateness. With characteristic decisiveness, he moved to position himself at the head of the central table. His lean frame straightened from its typical lounging posture into something that commanded attention without demanding it.

"Good people of Emberfield," he called, his voice carrying the particular resonance that had served him well as a performer among the nomadic Stone Caravan. "I've been considering what would most offend the entities who sought to destroy yesterday's joyful union."

This unexpected opening caught attention more effectively than conventional comfort might have achieved. Curiosity replaced uncertainty in many expressions as Theryn continued, his green-gold eyes scanning the gathering with deliberate inclusiveness.

"After careful consideration," he announced, lips curving into a smile, "I've concluded it would be nothing short of an affront to the gods and the Loom itself if we didn't feast on every last morsel of this wedding food and drink until we can't see straight."

The statement hung as its implications registered—permission granted through unlikely framing, necessity recast as rebellion, consumption transformed from potential disrespect to an act of defiance. Then laughter erupted from an unexpected source—a grandmother whose grandchild had been among those injured, her weathered face creasing with genuine mirth, with no denial of pain, but refused to be limited by it.

"The boy speaks truth," she declared, raising gnarled hands toward gathered neighbors. "My Teller would be first in line were his leg not broken. Shall we waste good food while those shadow-fire bastards watch and gloat?"

Her crude language—shocking from a respected elder whose piety had never before permitted such expression—broke remaining tension like ice cracking beneath spring sun. A cheer rose from gathered villagers, the sound gaining strength as it rippled outward through the square. Bodies surged toward tables with newfound purpose, hands reaching for serving utensils and plates, voices calling for ale casks to be breached.

"That's the spirit!" Theryn encouraged, lifting the nearest tankard in salute that encompassed the entire gathering. "Eat in their faces! Drink to their disappointment! Show those divine puppeteers that Emberfield knows how to survive, rebuild, and live despite their best efforts!"

The feast exploded into being with enthusiasm that might have seemed inappropriate to an outside observer unaware of the context. Platters circulated with increasing speed, portions growing more generous with each passing. Roasted meats—haunches of venison, whole chickens stuffed with winter herbs, rabbits wrapped in preserved grape leaves—disappeared onto plates that somehow never remained empty for long. Bowls of root vegetables glazed with honey and spices passed from hand to hand, their contents diminishing while conversation volume increased proportionally.

Children led the transformation from solemn aftermath to determined celebration, their natural resilience asserting itself through laughter that began cautiously but soon gained confidence. A boy with a bandaged arm demonstrated how effectively his uninjured hand could still transport honey cake from plate to mouth, crumbs dusting his shirt as evidence of successful technique. Two girls turned serving into an impromptu dance, balancing bread baskets while executing simple spinning steps that honored traditional wedding movements while adapting them to the current circumstance.

Adults followed more slowly but with growing conviction, the simple act of sharing food, weaving community bonds that had been tested but not severed by yesterday's violence. An elder raised a toast to "stubborn Emberfield stock—too tough for gods to chew properly," which prompted a round of appreciative laughter followed by enthusiastic emptying of tankards. A young woman with a fresh-stitched wound across her cheek insisted on serving others before herself, her injured face set in an expression that transformed pain into purpose.

Amid this emerging celebration, Theryn moved through the crowd with practiced ease, refilling cups before emptiness could suggest limitation, directing serving dishes toward those who looked hesitant, using humor to disarm lingering grief without denying its existence. His performance contained no artifice despite its deliberate nature—the genuine care behind his actions was evident in how precisely he calibrated each interaction to the specific needs of its recipient.

When momentarily unobserved, his expression revealed a deeper understanding of what he had orchestrated. "We need more of this," he murmured to himself, genuine sincerity replacing his typical ironic detachment. His fingers closed around the tankard handle with unusual firmness, as if anchoring himself to physical reality against the weightless drift of excessive contemplation. The slight smile tugging at his lips contained no mockery, only appreciation for how readily these villagers embraced life despite an intimate introduction to its fragility.

The companions gradually joined the celebration, each in a manner consistent with their individual nature. Bran approached with scholarly caution, plate balanced precisely as he cataloged nutritional content with automatic mental calculations before allowing himself to experience flavors directly. Alira moved among tables with healer's awareness, identifying those who needed encouragement to eat, occasionally touching a shoulder or hand in a gesture that conveyed permission to nourish body despite wounded spirit.

Edrin and Lysera entered the square together. The love glyph at Edrin's wrist pulsed gently as they approached the gathering, its subtle illumination visible to those with eyes to see such things. Villagers made space for them at the central table without fanfare or awkwardness—not as the center of an abandoned wedding but as community members sharing a common meal.

"Your ridiculous friend has interesting ideas about theological rebellion," Edrin observed to Theryn as he accepted the offered tankard. The words might have seemed critical without the warmth beneath them, the genuine appreciation evident in how tension had eased from his shoulders.

"The most profound defiance often wears a simple disguise," Theryn replied with uncharacteristic directness. "Breaking bread together when forces of destruction expected you to be broken yourselves—what could possibly irritate divine meddlers more thoroughly?"

The celebration continued into the afternoon, sunlight strengthening as if nature itself approved of this transformation from mourning to determined joy. What had begun as a practical distribution of food that would otherwise spoil evolved into a genuine communal experience.

Each bite consumed, each drink shared, each laugh that rose from tables represented a choice to continue living fully despite awareness of life's precarious nature.

In this simple feast, arranged on winter-cold cobblestones beneath open sky rather than within a decorated hall, Emberfield found expression of resilience more authentic than formal ceremony might have provided. They ate and drank not to forget what had occurred but to affirm that what remained—community, connection, daily pleasures—deserved protection and celebration precisely because of its vulnerability, not despite it.

During the feast, Edrin whispered to Lysera, "We could just make a small cottage for ourselves here and let the rest of this world rot away if it wants."

Theryn, as always, appeared behind him. "You won't do that," he whispered back, just as softly as Edrin had spoken.

"I would."

Theryn's smile never dimmed, "No, because the cottage that you want to build is in this world. You want to protect it. You want to shape it. You want to..."

Edrin interrupted, "I want to burn it all down. I want to wake up the gods and tell them, 'hands off.' I want to see the loom dismantled, and I want to see all creatures in this world do what they will without being told by a divine council what is right and what is wrong."

Lysera shook her head, smiling, "You have to build that cottage first, Edrin, otherwise you can't burn it all down. Very soon. You may have to learn how to be more than just a forgeman." Her eyes glazed over, a vision coming upon her, but it didn't seem to be impacting her the way that her other visions had lately. "You may have to learn how to fight, or do magic, or even become divine."

"Flame that!" Edrin exclaimed aloud, startling some of the elders near him. He looked sheepish and lowered his voice. "If I have to turn into something like Malrik to do all of that, it isn't worth it."

"Well," continued Theryn, as amiably as ever, "maybe if I am there… You won't need to go so far as that." Theryn's voice dropped to a low whisper, "These good people are going to get good and happy, and they should sleep in. This gives you two time to get to the top of that hill," he pointed at the nearby copse of trees at the top of a nearby hill, "At morning light's first shine, go and marry, thee and thine."

Edrin couldn't help it. He smiled a rare, full smile and slightly nodded his head. It felt good to smile.

At dawn the next morning, they did what Theryn told them to do.

Somewhere beneath Threadmere – Night (Morday, 5 Concord Days, TT 1215)

In a tower chamber that existed at uncomfortable angles to ordinary reality, the mirror mage stood surrounded by mirrors that reflected neither truth nor falsehood but possibilities in between. The circular room housed dozens of reflective surfaces in varying sizes and shapes—some tall enough to capture full human form, others small as a thumbnail but containing vistas that should have required much greater scope. None hung perfectly straight upon the walls of black stone; each tilted at a precise degree that created subtle disorientation in viewers unaccustomed to such deliberate perceptual manipulation. Illumination came from sources that cast no shadows—pale blue flames that burned in suspended crystal vessels, their light possessing a quality that seemed to reveal surfaces while concealing depths. While this wasn't her sanctum, she used it to keep her eyes, ears, and mind on what was going on out in the world while she stayed in the shadows.

The sorceress moved with liquid grace between these reflective sentinels, her fingers occasionally brushing against frames of silver, obsidian, and materials less easily defined. Her hair—shifting constantly between violet, silver, and midnight blue—wove with subtle autonomy. Before the central mirror—an oval of polished thunderglass suspended within a ring of blackened silver—she finally paused, extending one perfect hand toward its surface with deliberate intent rather than casual contact.

"Show me," she commanded, voice composed of multiple harmonics that created an impression of several speakers sharing a single throat. "The smith and his companions. Emberfield. Now."

The mirror's surface rippled like disturbed water, its reflective property temporarily surrendering to translucence before resolving into a clear image of a cobblestoned square where villagers had gathered for an improvised feast. Unlike common scrying that provided a distant, fixed perspective, this viewing adjusted continually—focusing briefly on Edrin's hands as he raised a tankard in toast, then widening to capture Lysera beside him, then shifting to observe Bran's methodical documentation of food distribution patterns. The vision possessed the quality of presence rather than mere observation, as if part of the ritualist's consciousness had been extended into physical location far from her tower sanctuary.

Her fractured eyes—each iris composed of countless mirror shards that reflected miniature versions of whatever they beheld—narrowed as she absorbed details invisible to ordinary sight. Beyond physical actions and material surroundings, she perceived energetic patterns flowing between companions—threads of connection whose colors and intensities revealed emotional states and bonds stronger than mere alliance of convenience.

"Fascinating resilience," she murmured, head tilting slightly as observation continued. "The vessel's attack strengthened rather than severed their connections. How predictably contrary to divine intention."

With a subtle gesture of her free hand, the mirror's focus shifted to the low table where her other hand rested upon a spread of parchments. These documents—some bearing official seals of authority, others marked with glyphs of secrecy and containing information never meant for preservation in written form—

created a complex mosaic of intelligence gathered from sources both mundane and supernatural. Her fingers traced patterns across reports detailing Malrik's attack, witness testimonies, astronomical calculations, and magical measurements taken from a distance by agents whose identities remained concealed even in private documentation.

"The vessel performed adequately despite structural limitations," she noted to an apparently empty chamber, though something in the furthest mirror shifted slightly at her words—a presence observing the observer, permitted such access only within carefully defined parameters. "Significant damage inflicted, psychological impact achieved, divine energies properly displayed." Her finger tapped against parchment bearing a detailed accounting of injuries sustained during the attack. "Though the ultimate objective remains unfulfilled."

As she spoke, her other hand rose to the pendant suspended at her throat—jagged mirror-shard whose edges appeared sharp enough to draw blood, though they never marked her perfect skin. The concealment sigil etched upon its surface pulsed with a subtle rhythm that mirrored neither her heartbeat nor any natural biological cycle but rather cosmic frequency beyond mortal measurement. This protective enchantment ensured her observation remained undetected even by Lysera's heightened seer perception—a veil between watcher and watched that consumed significant magical resources to maintain yet represented necessary investment given the stakes involved.

"Convergence calculations require adjustment," the mirror witch continued, turning from the central mirror toward the table covered in star charts and esoteric diagrams. With precise movements that suggested both artistic sensibility and mathematical rigor, she modified constellation alignments, adjusted planetary positions, and recalculated energy flows based on the most recent observations. Each adjustment caused ripples through the entire system—a change in a single variable necessitating comprehensive reconsideration of all related factors.

"Pyrrhost ascends during Emberturn's final phase," she murmured, fingers dancing across parchment with inhuman speed and accuracy. "Umbryl reaches apogee three days later during Mistbloom's beginning. Suboptimal alignment for direct intervention but serviceable for subtle manipulation." She made a notation in script not used in mortal communication for centuries, its angular forms containing conceptual compression that allowed complex ideas to occupy minimal space. "The vessel will require substantial restoration before deployment becomes viable. Fourteen days minimum, twenty-three optimal."

Returning to the central mirror, the watcher adjusted its focus with a delicate gesture, causing the image to center upon Edrin and Lysera—specifically their clasped hands where love glyph pulsed with golden light visible only to those capable of perceiving magical emanations. Her fractured eyes reflected dozens of miniature versions of this connection, each tiny image examining the bond from a slightly different angle of perception.

"Their bond makes them dangerous," she whispered, words emerging with intensity that belied her outward calm. "The glyph manifested spontaneously rather than through ritual invocation—evidence of Loom recognition beyond ordinary pattern." Her finger traced the outline of golden light upon the mirror surface, careful not to actually touch the image itself lest contact create a connection that might be detected. "But love remains the easiest thread to cut—not through direct assault but by redirecting its natural tensions until it frays from within."

The mirror's focus shifted again at her unspoken command, capturing an image of Bran's almanac pages—complex calculations and astronomical notations recorded in his meticulously precise hand. The veiled adept's expression revealed neither surprise nor concern at this evidence of analysis running parallel to her own, merely a cool assessment of the competitor's methodology.

"Impressive pattern recognition," she acknowledged, "though his understanding remains fundamentally limited by mortal perspective. He perceives shadows cast by greater movements but cannot comprehend the bodies that cast them." Her lips curved in a smile, a mere appreciation of ironic limitation. "Still, his calculations approach dangerous proximity to actual convergence dates. Adjustment of their trajectory becomes increasingly critical."

Moving to a different table where a map had been pinned with obsidian weights at each corner, the mirror-worker began marking new coordinates with ink that shifted between silver and shadow depending on the angle of observation. Each notation corresponded to a location where ley-line energies converged with astronomical alignments to create nodes of power suitable for specific types of magical working. Her markings created a complex geometric pattern when connected—not random points but a carefully structured network that, when activated in proper sequence, would funnel energy toward a predetermined outcome.

"The next moves require subtlety," she stated to mirror-presence that observed without commenting. "Not direct confrontation but separation—divide their strength, isolate their weaknesses, introduce elements that create internal conflict rather than external threat." Her hand moved to a specific location on the map—a mountain pass where three ley-lines intersected beneath a constellation representing divine judgment. "Here. When Pyrrhos and Umbryos reach momentary harmony in twenty-seven days. The vessel will be ready by then, and their path will lead them directly into position."

The central mirror's surface began clouding as the practitioner turned away, her attention shifting from observation to implementation. As the connection faded, her silhouette reflected in multiple surfaces simultaneously, each mirror capturing a slightly different version of her form—some showing a woman of cold beauty, others revealing something less human beneath a perfect exterior, still others displaying a fragmented collection of aspects that suggested identity composed of carefully assembled rather than naturally integrated elements.

These reflections moved with subtle differences from one another and from the mirror mage herself, creating an impression of being not merely images but possible versions existing in parallel—potential manifestations of an entity whose true nature remained dispersed across multiple planes of existence rather than contained within a single physical form. Each reflection performed a slightly different action—consulting a different document, making an alternative calculation, considering a separate strategy—all serving a unified purpose while maintaining distinct approaches.

In the final moment before scrying connection was severed completely, one mirror briefly revealed the illusionist's true expression beneath a carefully maintained mask of control—a combination of hunger and calculation, ambition and patience, desire and restraint, transforming beautiful features into something altogether more dangerous than mere malevolence. This glimpse—too brief to be fully comprehended yet sufficient to suggest depths beyond simple villainy—contained the essence of her nature: not opponent but architect, not servant but manipulator, not merely player in divine game but one who sought to rewrite its fundamental rules.

Then the connection collapsed entirely, leaving only an empty tower chamber filled with mirrors that once again reflected fragmentary versions of present reality rather than distant observation—waiting tools prepared for future use by hands that understood precisely how to wield them against those who believed themselves beyond such manipulation.

All of her forms, her puppets, Malrik, Councilor Vaelis, Mirror Mage, Witch of reflections, her multitude of images went out into the world to see what they could do to turn things to their master's will, and her face smiled, but those shardlike eyes did not waver.

And they responded to her will, as one, "Yes, Mistress Mirrathar."

The End

The End

Made in the USA
Monee, IL
13 September 2026

62981859R10233